Applied Genetic Programming and Machine Learning

The CRC Press

International Series on Computational Intelligence

Series Editor
L.C. Jain, Ph.D., M.E., B.E. (Hons), Fellow I.E. (Australia)

L.C. Jain, R.P. Johnson, Y. Takefuji, and L.A. Zadeh
Knowledge-Based Intelligent Techniques in Industry

L.C. Jain and C.W. de Silva
**Intelligent Adaptive Control: Industrial Applications in the
Applied Computational Intelligence Set**

L.C. Jain and N.M. Martin
**Fusion of Neural Networks, Fuzzy Systems, and Genetic Algorithms:
Industrial Applications**

H.-N. Teodorescu, A. Kandel, and L.C. Jain
Fuzzy and Neuro-Fuzzy Systems in Medicine

C.L. Karr and L.M. Freeman
Industrial Applications of Genetic Algorithms

L.C. Jain and B. Lazzerini
Knowledge-Based Intelligent Techniques in Character Recognition

L.C. Jain and V. Vemuri
Industrial Applications of Neural Networks

H.-N. Teodorescu, A. Kandel, and L.C. Jain
Soft Computing in Human-Related Sciences

B. Lazzerini, D. Dumitrescu, L.C. Jain, and A. Dumitrescu
Evolutionary Computing and Applications

B. Lazzerini, D. Dumitrescu, and L.C. Jain
Fuzzy Sets and Their Application to Clustering and Training

L.C. Jain, U. Halici, I. Hayashi, S.B. Lee, and S. Tsutsui
Intelligent Biometric Techniques in Fingerprint and Face Recognition

Z. Chen
Computational Intelligence for Decision Support

L.C. Jain
Evolution of Engineering and Information Systems and Their Applications

CRC Press International Series on Computational Intelligence

Applied Genetic Programming and Machine Learning

Hitoshi Iba

Topon Kumar Paul

Yoshihiko Hasegawa

CRC Press
Taylor & Francis Group
Boca Raton London New York

CRC Press is an imprint of the
Taylor & Francis Group, an **informa** business

CRC Press
Taylor & Francis Group
6000 Broken Sound Parkway NW, Suite 300
Boca Raton, FL 33487-2742

First issued in paperback 2019

© 2010 by Taylor and Francis Group, LLC
CRC Press is an imprint of Taylor & Francis Group, an Informa business

No claim to original U.S. Government works

ISBN-13: 978-1-4398-0369-1 (hbk)
ISBN-13: 978-0-367-38527-9 (pbk)

Visit the Taylor & Francis Web site at
http://www.taylorandfrancis.com

and the CRC Press Web site at
http://www.crcpress.com

Contents

List of Tables

List of Figures

Preface

This book delivers theoretical and practical knowledge on extension of Genetic Programming (GP) for practical applications. It provides a methodology for integrating Genetic Programming and machine-learning techniques. The development of such tools contributes to the establishment of a more robust evolutionary framework when addressing tasks from such areas as chaotic time-series prediction, system identification, financial forecasting, classification, and data mining.

The overall goals of this textbook are:

- offer a comprehensive overview of principles of GP and machine-learning techniques;

- dissect the computational and design principles behind the integration of GP and machine-learning techniques towards designing new and better tools;

- understand how to apply the proposed approach in the real-world areas;

- provide listings of available source codes and GUI (Graphical User Interface) systems, as practical applications for the readers to grasp the concepts in a proper manner;

- offer a perspective on current Evolutionary Computation research;

The key idea is to integrate GP with machine-learning techniques in the following ways:

1. GP's tree-structured representation can be flexibly tailored to the data in the form of a PNN (Polynomial Neural Network). The PNN model is frequently used with contemporary neural network methods for weight training and pruning, as an alternative learning method for statistical data analysis. Search for optimal tree-like topologies can be organized using GP. Search in the weight space can be implemented using a multiple-regression method.

2. The ensemble technique is employed to improve the classification performance of GP. GP multiple-voting technique is proposed and compared with the traditional machine-learning methods, i.e., AdaBoost and Bagging method.

3. Probabilistic learning technique is used to enhance the program's evolution. GP is integrated with Bayesian inference method and grammatical inference approach. This is based on the idea of EDA (Estimation of Distribution Algorithm), which has been frequently used in the GA (Genetic Algorithm) community. We describe two types of approach in GP-EDAs: prototype tree based methods and probabilistic context free grammar (PCFG) based methods.

The emphasis of this book is GP's applicability to real-world tasks. Empirical examples from these real-world data are given to show how to preprocess the data before learning. Tasks from the following application areas are studied:

- Financial data prediction, e.g., stock market pricing and FX (Foreign eXchange) data;

- Day-trading rule development;

- Biomarker selection in bioinformatics;

- Defaulter detection from financial data;

The empirical investigations demonstrate that the proposed approaches are successful when solving real-world tasks.

Covering the three fields: evolutionary computation, statistical inference method, and machine-learning orient the book to a large audience of researchers and practitioners from the computer sciences. Researchers in GP will study how to use machine-learning techniques, how to make learning operators that efficiently sample the search space, how to navigate the search process through the design of objective fitness functions, and how to examine the search performance of the evolutionary system. The possibility to use reliable means for observing the search behavior of GP systems is one of their essential advantages.

Students will find this book useful for studying machine-learning technique, probabilistic learning, and GP. Undergraduate students will learn how to design and implement the basic mechanisms of a GP system, including the selection scheme, the crossover and mutation learning operators. Postgraduate students will study advanced topics such as improving the search control of GP systems, and tools for the examination of their search performance.

The book offers statisticians a shift in focus from the standard linear models towards highly non-linear models that can be inferred by contemporary learning approaches. Researchers in statistical learning will read about alternative probabilistic search algorithms that discover the model architecture, and find accurate polynomial weights by neural network training techniques. It is important to note that these researchers may be inspired by the possibility to easily interpret the discovered models, and to carry out statistical diagnosis of these models by standard statistical means.

We also provide the essential knowledge about data classification using the ensemble of genetic programming rules, which is very accessible to readers of different levels of expertise. Those who are novice in the field of classification and feature selection will find here a sound base for their applications and further investigations and discussions. The bio-informaticians and the medical scientists will find a new machine-learning tool that can be used to support the prediction of cancer class using microarray data and for the identification of the potential biomarker genes. Finally, the various evaluation techniques discussed in this chapter will assist those who are interested in developing a reliable prediction system for the prediction of a rare event using imbalanced data.

The developments in this book aim to facilitate the research of extended GP frameworks with the integration of several machine-learning schemes. The empirical studies are mainly taken from practical fields, e.g., system identification, financial engineering and bioinformatics. The interest in such natural problems is that many everyday tasks actually fall in this category. Experimental results shown in the book suggest that the proposed methodology can be useful in practical inductive problem solving.

Iba, H., Paul, T. K., Hasegawa, Y.,
Tokyo, Japan, October, 2008.

Acknowledgments

To all those wonderful people we owe a deep sense of gratitude especially now that this book project has been completed. Especially, we acknowledge the pleasant research atmosphere created by colleagues and students from the research laboratory associated with Graduate School of Frontier Sciences at the University of Tokyo.

The first author, Dr. Hitoshi Iba, is grateful to his previous group at Electro-Technical Laboratory (ETL), where he worked for ten years, and to his current colleagues at Graduate School of Engineering of the University of Tokyo. Particular thanks are due to Dr. Hirochika Inoue and Dr. Taisuke Sato for providing precious comments and advice on numerous occasions. He also wishes to thank Dr. Nikolay Nikolaev for the collaboration with STROGANOFF project. And last, but not least, he would like to thank his wife Yumiko and his sons and daughter Kohki, Hirono and Hiroto, for their patience and assistance.

The second author, Dr. Topon Kumar Paul, owes a debt of gratitude to Toshiba Corporation, where he has been working as a researcher since April 2007, for all kinds of financial support and encouragement for writing the book. He is indebted to all the colleagues of System Engineering Laboratory at Corporate Research & Development Center of Toshiba Corporation for their many suggestions and constant support during the book project. He is especially grateful to Dr. Juan Liu, ex-member of IBA Laboratory of the

University of Tokyo, for her work that motivated him to conduct research on classification and feature selection. Finally, he would like to thank his wife Banani and two daughters Tanusree and Bhagyashree for their patience and inspiration.

The third author, Dr. Yoshihiko Hasegawa, is grateful to Dr. Kohsuke Yanai for discussing his works and to students at Iba laboratory for giving precious comments on his chapter. He would also like to express special appreciation to his parents for supporting his research.

Chapter 1

Introduction

> Science is − and how else can I say it? − when it plays with interesting ideas, examines their implications, and recognizes that old information may be explained in surprisingly new ways. Evolutionary theory is now enjoying this uncommon vigor. (Stephen Jay Gould, "Hen's Teeth and Horse's Toes", W. W. Norton & Company, 1994)

"Why are the peacock's feathers so incredibly beautiful?"

"Why did the giraffe's neck become so long?"

"If a worker bee cannot have any offspring of its own, why does it work so hard to serve the queen bee?"

If we make a serious effort to answer these mysteries, we realize that we are solving one of the problems of optimization for each species, i.e., the process of evolution of species. It is the objective of the evolutionary method to exploit this concept to establish an effective computing system (an evolutionary system). Evolutionary computation (EC) attempts to "borrow" Nature's methods of problem solving and has been widely applied to find solutions to optimization problems, to automatically synthesize programs, and to accomplish other AI (Artificial Intelligence) tasks for the sake of effective learning and formation of hypotheses. EC imitates the evolutionary mechanisms of living organisms to create, to combine, and to select data structures. This method takes the mechanisms of reproduction and selection as a model and are widely applied in engineering fields. This is called "the evolutionary approach."

The data handled by EC has a two-layer structure, phenotype (PTYPE) and genotype (GTYPE, or "genetic code", corresponding to the chromosomes in a cell). The GTYPE is the analogue of the gene; it is a set of low-level, local rules that are the object of the action of genetic operators (see Fig. 1.1). The PTYPE represents the phenotype and expresses the global behaviors and structures accompanying the expression of the GTYPE in the environment. The fitness value is uniquely determined by the PTYPE within the context of the environment; therefore, fitness selection is dependent on PTYPE. Table 1.1 shows the analogues in EC for biological functions.

In EC, a population is made up of multiple individuals. Each individual has its own GTYPE or genetic code, and its fitness value is set with respect to the expressed PTYPE. These individuals reproduce or recombine to create a succeeding generation. This process is controlled to increase the number of

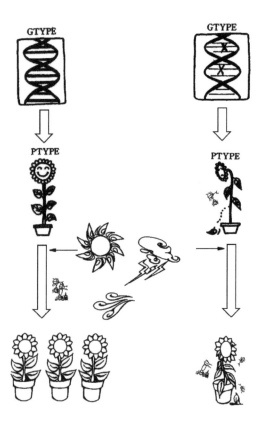

FIGURE 1.1: GTYPE and PTYPE.

Table 1.1: Analogy with biological organisms.

	GTYPE Genotype	PTYPE Phenotype	Fitness value
Biology	Genes	Proteins produced, Expressed functionalities	Ease of survival, Number of offspring
GP	Tree structure	Programs, Conceptual structures	Performance measures
GA ES EP	Character strings, Numerical values, Automata	Functional values, Generated strings	Closeness to optimal solution

likely offspring of individuals in keeping with the high quality of their fitness and to suppress reproduction by individuals with low fitness (the biological term for this is "selection", i.e., the survival of the fittest). Genetic operators are applied to cause crossover, mutation, and other patterns of transmission of GTYPE during the reproduction, to create the GTYPE of the succeeding generation. These various operators in EC recombine genes in ways mimicking mutation and other processes in biology. The frequency of application of these operators and the locations in GTYPE where they are applied are generally determined by means of a random process. It is intended that each succeeding generation contain individuals whose expressions are slightly varied from the previous generation and so that the fitness value be higher, in turn, raising the overall fitness of the population. In the same way, each succeeding generation gives birth to the next one, and as this process is repeated, the basic theory of EC is that the fitness of the overall population increases down the generations. Fig. 1.2 shows the illustrative image of EC, in which the fitness value of each individual is represented within the ellipse. The larger the value, the better the fitness in the figure.

Thus, we can summarize the basic flow of EC as follows. A group of individuals is designated as a set $M(t) = \{g_t(m)\}$ of genotype GTYPE at the t-th generation. The fitness value $u_t(m)$ of the phenotype PTYPE $p_t(m)$ for each individual $g_t(m)$ is then determined in the environment. The genetic operators are generally applied to the GTYPE with good (high) fitness values; as a result, these replace the GTYPE with bad (low) fitness values. This is the selection based on the fitness value, and the set of the next generation of GTYPEs $M(t+1) = \{g_{t+1}(m)\}$ is produced. This process is then repeated in the same way.

EC provides approaches for doing simultaneously global and local search [Nikolaev and Iba06]. The main evolutionary paradigms are: Genetic Algorithm (GA) [Holland75, Goldberg89], Genetic Programming (GP) [Koza92, Koza94, Koza *et al.*99, Koza *et al.*03, Banzhaf *et al.*98, Langdon and Poli02, Riolo and Worzel03, O'Reilly *et al.*05, Yu *et al.*06a], Evolution Strategies (ES) [Bäck96, Schwefel95, Bäck *et al.*00, Eiben and Smith03], and Evolutionary

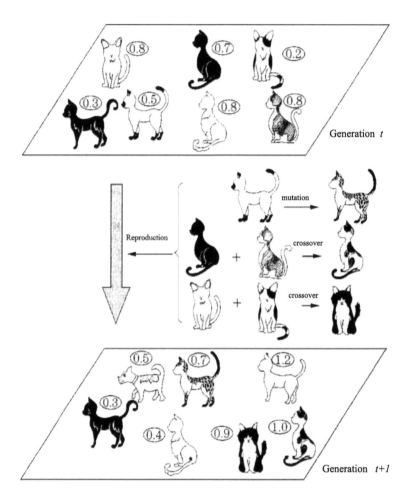

FIGURE 1.2: Evolutionary computation.

Programming (EP) [Fogel *et al.*66, Fogel99]. They have conducted probabilistic population-based search which is a powerful tool for broad exploration and local exploitation of the model space. The population-based strategy is an advantage over other global search algorithms such as simulated annealing [Kirkpatrick *et al.*83] and tabu search [Glover89], which works with only one hypothesis at a time, and over algorithms for local search [Atkeson *et al.*97] that perform only narrow examination of the search space. Their stochastic character is an advantage over the heuristic AI [Nilsson80, Nilsson98] and machine-learning algorithms [Mitchell97, Smirnov01] that also search with one hypothesis. The next chapter describes the basics of GP in detail.

Chapter 2

Genetic Programming

2.1 Introduction to Genetic Programming

The aim of Genetic Programming (GP) is to extend genetic forms from Genetic Algorithm (GA) to the expression of trees and graphs and to apply them to the synthesis of programs and the formation of hypotheses or concepts. Researchers are using GP to attempt to improve their software for the design of control systems and structures for robots.

The original concept of GP was conceived by John Koza of Stanford University and his associates [Koza92]. GP is one of the fields of evolutionary computation. When the concepts of GP are applied in AI (Artificial Intelligence), the processes of learning, hypothesis formation and problem solving are called "evolutionary learning" or Genetic-Based Machine Learning (GBML). This learning method is based on fitness and involves transformation of knowledge and elimination of unfit solutions by a process of selection to preserve appropriate solutions in the subsequent generation. It has much in common with classifier systems [Wilson87].

The procedures of GA are extended in GP in order to handle graph structures (in particular, tree structures). Tree structures are generally well described by S-expressions in LISP. Thus, it is quite common to handle LISP programs as "genes" in GP. As long as the user understands that the program is expressed in a tree format, then he or she should have little trouble reading a LISP program (the user should recall the principles of flow charts). The explanations below have been presented so as to be quickly understood by a reader who does not know LISP.

A tree is a graph with a structure as follows, incorporating no cycles:

More precisely, a tree is an acyclical connected graph, with one node defined as the root of the tree. A tree structure can be expressed as an expression with parentheses. The above tree would be written as follows:

```
(A (B)
  (C (D)))
```

In addition, the above can be simplified to the following expression:

```
(A B
  (C D))
```

This notation is called an "S-expression" in LISP. Hereinafter, a tree structure will be identified with its corresponding S-expression. The following terms will be used for the tree structure:

- Node: Symbolized with A, B, C, D, etc.

- Root: A

- Terminal node: B, D (also called a "terminal symbol" or "leaf node")

- Non-terminal node: A, C (also called a "non-terminal symbol" and an "argument of the S-expression")

- Child: From the viewpoint of A, nodes B and C are children (also, "arguments of function A")

- Parent: The parent of C is A

Other common phrases will also be used as convenient, including "number of children", "number of arguments", "grandchild", "descendant", and "ancestor." These are not explained here, as their meanings should be clear from the context.

The following genetic operators acting on the tree structure will be incorporated:

1. **Gmutation** Alteration of the node label

2. **Ginversion** Reordering of siblings

3. **Gcrossover** Exchange of a subtree

These are natural extensions of existing GA operators and act on sequences of bits. These operators are shown below in examples where they have been applied in LISP expression trees (S-expressions) (Fig. 2.1). The underlined portion of the statement is the expression that is acted upon:

Gmutation   Parent:$(+\ x\ \underline{y})$
                        $\Downarrow$
               Child:$(+\ x\ \underline{z})$

Ginversion   Parent:$(\text{progn}\ \underline{(\text{incf}\ x)\ (\text{setq}\ x\ 2)}\ (\text{print}\ x))$
                        $\Downarrow$
               Child:$(\text{progn}\ \underline{(\text{setq}\ x\ 2)\ (\text{incf}\ x)}\ (\text{print}\ x))$

Gmutation

Ginversion

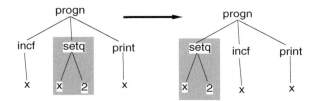

Gcrossover

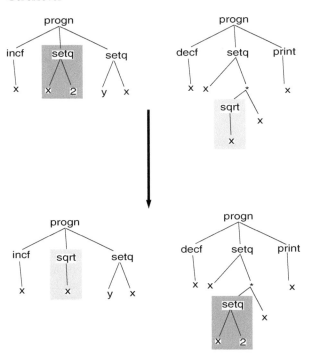

FIGURE 2.1: Genetic operators in GP.

Gcrossover   Parent₁:(progn (incf x) (setq x 2) (setq y x))
                        Parent₂:(progn (decf x) $\overline{\text{(setq } x \text{ (}* \text{ (sqrt } x) \text{ } x))}$ (print x))
                        ⇓
                        Child₁:(progn (incf x) (sqrt x) (setq y x))
                        Child₂:(progn (decf x) $\overline{\text{(setq } x \text{ (}* \text{ (setq } x \text{ 2) } x))}$ (print x))

Table 2.1 provides a summary of how the program was changed as a result of these operators. "progn" is a function acting on the arguments in the order of their presentation and returns the value of the final argument. The function "setq" sets the value of the first argument to the evaluated value of the second argument. It is apparent on examining this table that mutation has caused a slight change to the action of the program, and that crossover has caused replacement of the actions in parts of the programs of all of the parents. The actions of the genetic operators have produced programs that are individual children but that have inherited the characteristics of the parent programs.

More strictly, we use the following kinds of Gmutation: (Fig. 2.2).

1. Mutations that change a terminal node to a non-terminal node, corresponding to the creation of a subtree (Fig. 2.2(a)).

2. Mutations that change a terminal node to another terminal node, changing only the node label (Fig. 2.2(b)).

3. Mutations that change a non-terminal node to a terminal node, corresponding to the deletion of a subtree (Fig. 2.2(c)).

4. Mutations that change a non-terminal node to another non-terminal node.

   Case 1   The new non-terminal node has the same number of children as the old non-terminal node (Fig. 2.2(d)).
            ⇒ Only the node label is changed.

   Case 2   The new non-terminal node has a different number of children from the old non-terminal node (Fig. 2.2(e)).
            ⇒ A subtree is created or deleted.

The application of the above genetic operators is controlled stochastically.

Except for the aspect that the genetic operator acts on the structural representation, the GP employs a standard GA process (see Fig. 2.3). The original program (the structural representation) changes a little at a time under the action of the genetic operators illustrated in Table 2.1. The GP searches for the desired program by the same process of selection operations.

We will describe the selection techniques used in a GP search. The fundamental principle of GP is that candidate parents are selected by highest (best) fitness value to produce large numbers of offspring. There are several ways to accomplish this; the following subsections explain the most popular ones.

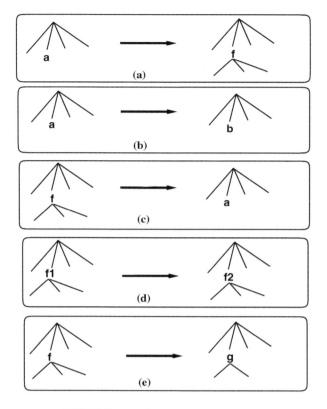

FIGURE 2.2: Gmutation operators.

Table 2.1: Program changes due to GP operators.

Operator	Program before operation	Program after operation
Mutation	Add x and y.	Add x and z.
Inversion	1. Add 1 to x. 2. Set $x = 2$ 3. Print $x(=2)$ and return 2.	1. Set $x = 2$. 2. Add 1 to x. 3. Print $x(=3)$ and return 3.
Crossover	Parent$_1$: 1. Add 1 to x. 2. Set $x = 2$. 3. Set $y = x(= 2)$ and return 2. Parent$_2$: 1. Subtract 1 from x. 2. Set $x = \sqrt{x} \times x$. 3. Print x and return the value.	Child$_1$: 1. Add 1 to x. 2. Take square root of x. 3. Set $y = x$ and return the value. Child$_2$: 1. Subtract 1 from x. 2. Set $x = 2$ and its value $(=2)$ is multiplied by $x(=2)$. The result value $(=4)$ is set to x again. 3. Print $x(=4)$ and return 4.

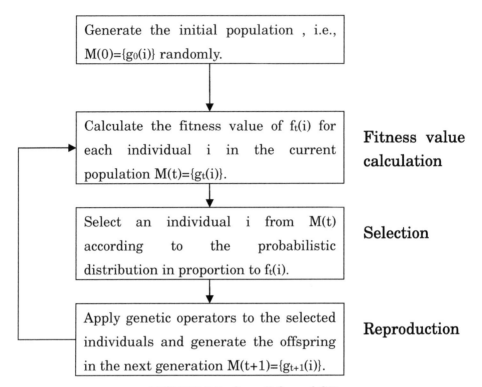

FIGURE 2.3: Overall flow of GP.

2.1.1   Roulette-Wheel Selection

This is a method in which a selection is made with probability proportional to the fitness value. The method is also called a fitness-proportionate selection. The simplest version of this method is weighted roulette-wheel selection. A roulette-wheel whose sectors' widths are proportional to the fitness values is "spun", and the sector where the "ball" falls is selected. For example, let us consider the case where

$$f_1 = 1.0$$
$$f_2 = 2.0$$
$$f_3 = 0.5$$
$$f_4 = 3.0$$
$$f_5 = 3.5$$

The above selection on a weighted roulette-wheel would be expressed as follows. First, we calculate the summation of the fitness values, i.e.,

$$f_1 + f_2 + \cdots + f_5 = 10.0. \tag{2.1}$$

Next, we obtain uniformly distributed random numbers between 0 and 10; f_i is then selected as follows:

When the random number is in $[0.0, 1.0)$, f_1 is selected.
When the random number is in $[1.0, 3.0)$, f_2 is selected.
When the random number is in $[3.0, 3.5)$, f_3 is selected.
When the random number is in $[3.5, 6.5)$, f_4 is selected.
When the random number is in $[6.5, 10.0]$, f_5 is selected.

For example, if the random numbers are

$$1.3, \quad 5.5, \quad 8.5, \quad 4.5, \quad 7.5, \tag{2.2}$$

then this sequence is selected:

$$f_2, \quad f_4, \quad f_5, \quad f_4, \quad f_5. \tag{2.3}$$

This proceeds until a sequence with a number of members equaling the population size (n individuals) has been picked out.

Let us state the roulette-wheel selection method more formally. When the fitness values for n individuals have been determined as f_1, f_2, $\cdots$, f_n, the probability p_i that the i-th individual of the population will be selected is given by

$$p_i = \frac{f_i}{\sum f_i}. \tag{2.4}$$

Thus, the expected number of offspring born to the i-th individual is

$$np_i = \frac{nf_i}{\sum f_i} = \frac{f_i}{\sum f_i/n} = \frac{f_i}{f_{avg}}, \tag{2.5}$$

where f_{avg} is the mean fitness value $f_{avg} = \sum f_i/n$. It should be noted that the mean number of offspring for members of the population is

$$n \times p_{avg} = \frac{f_{avg}}{f_{avg}} = 1, \tag{2.6}$$

i.e., the average individual leaves 1 individual as offspring in the succeeding generation.

2.1.2 Greedy Over-Selection Method

Greedy over-selection [Koza92] is usually used when population is large, e.g., more than 1000 individuals. In this method, individuals are selected according to their fitness values. However, this method biases selection towards the highest performers.

To reduce the number of generations required for a GP run, Koza made use of the following strategy:

1. Using the fitness values, the population is divided into two groups. Group I includes the top 20% of individuals while Group II contains the remaining 80%.

2. Individuals are selected from Group I 50% of the time. The selection method inside a group is fitness-proportionate.

The percentages of the above strategies are user-defined parameters.

2.1.3   Ranking Selection Method

The ranking selection takes the individuals according to their arrangement in the population in increasing or decreasing order of fitness. The best individual is assigned rank one and the worst is given rank equal to the population size. The linear ranking scheme determines the selection probability of an individual with the following function [Nikolaev and Iba06]:

$$\Pr(\mathcal{G}_i^\tau) = \frac{1}{\pi}\left(\alpha + (\beta - \alpha)\frac{rank(\mathcal{G}_i^\tau) - 1}{\pi - 1}\right) \tag{2.7}$$

where $\mathcal{G}_i^\tau$ is the i-th genetic program at generation τ, $rank(\mathcal{G}_i^\tau)$ is its rank in the population, π is the population size, α is the proportion for selecting the worst individual, and β is the proportional for selecting the best individual. When the conditions: $\alpha + \beta = 2$ and $1 \leq \alpha \leq 2$ are satisfied, the best individual will produce no more than twice offspring than the population average.

There are several modifications: linear ranking, non-linear ranking, and exponential ranking. The linear ranking selection has been applied to GP [Koza92] using a factor r_m for the parameters α and β that simulates the function gradient as follows: $\alpha = 2/(r_m + 1)$, $\beta = 2r_m/(r_m + 1)$.

2.1.4   Truncation Selection Method

A similar selection scheme is the uniform ranking, known also as truncation selection [Mühlenbein and Schlierkamp95]. This scheme chooses for reproduction only from the top individuals according to their rank in the population, and these top individuals have the same chance to be taken. The fraction of the top individuals is defined by a rank threshold μ. The selection probability Pr of the truncation selection scheme is:

$$\Pr(\mathcal{G}_i^\tau) = \begin{cases} 1/\mu, & 1 \leq rank(\mathcal{G}_i^\tau) \leq \mu \\ 0, & \text{otherwise} \end{cases} \tag{2.8}$$

where $rank(\mathcal{G}_i^\tau)$ is the rank of the i-th individual $\mathcal{G}_i^\tau$ in the population at generation τ [Nikolaev and Iba06]. The fitnesses are used only for ordering of the population $f(\mathcal{G}_1^\tau) \leq f(\mathcal{G}_2^\tau) \leq \cdots \leq f(\mathcal{G}_\pi^\tau)$, and they do not participate in the selection. The advantage of truncation selection is that it is not affected directly neither by the fitnesses nor by the scaled adjusted values.

In estimation of distribution algorithm or estimation of distribution programming based methods, all the top-ranked individuals are used in estimation of probability distribution of the parameters (see Section 5.2 for details).

2.1.5   Tournament Selection Method

In the tournament selection method, a set of individuals (its size is designated as S_t) is selected at random from the population and the best of these is selected. This process is performed until a number of individuals corresponding to the size of the original population is obtained. Let us illustrate this process with the same example as before:

$$f_1 \quad 1.0$$
$$f_2 \quad 2.0$$
$$f_3 \quad 0.5$$
$$f_4 \quad 3.0$$
$$f_5 \quad 3.5$$

Let us suppose that $S_t = 3$ and there were five tournaments in which the following individuals were selected:

1st tournament $f_1 f_2 f_3$

2nd tournament $f_3 f_4 f_1$

3rd tournament $f_1 f_5 f_2$

4th tournament $f_2 f_4 f_1$

5th tournament $f_5 f_2 f_4$

The "winner", or selected individual, in each tournament was f_2, f_4, f_5, f_4, and f_5. There are many variations of this method with different tournament sizes (S_t).

2.1.6   Elite Strategy

In the previous methods, the candidate parents are always chosen stochastically, and the best individuals do not necessarily reappear in the succeeding generation. Even if they are left as candidate parents in the next generation, they may be randomly selected for mutations or crossover. Therefore, it is not guaranteed that the population will become more and more successful with the passing of generations.

In the "Elite strategy", the best individual is (or some of the most successful individuals are) "cloned" or retained from each generation to the next. Crossover and mutation are not applied to these "elite" individuals, i.e., they are simply copied. This guarantees that if the fitness function remains the same, the success of a generation will be, at least, no worse than that of the preceding generation. The share of successful individuals in a generation that should be left in the simulation can be a designated parameter.

2.1.7　Generation of Initial Population

Initial population of individuals is created by random valid compositions of functions and terminals. For any node of the tree representing an S-expression, either a function or a terminal is chosen randomly with the restriction on the depth and the semantic of a rule. The pseudocode of creating a rule in "grow mode" is as follows:

```
GenerateRule(Tree t, Integer depth)
BEGIN
 IF (depth<1) THEN
    return;
 ELSEIF (depth=1) THEN /*only terminal*/
  BEGIN
   t.value=SelectTerminalRandomly();
   t.left=null; t.right=null;
   return;
  END
 ELSE
  BEGIN
    node=SelectNodeRandomly();
    IF (node is a terminal) THEN
     BEGIN
      t.value=node;
      t.left=null;
      t.right=null;
      return;
     END
    ELSEIF (node is a unary function) THEN
     BEGIN
      t.value=node;
      t.right=null;
      t.left=new Tree();
      GenerateRule(t.left,depth-1);
     END
    ELSE /*a binary function*/
     BEGIN
      t.value=node;
      t.left=new Tree();
      t.right=new Tree();
      GenerateRule(t.left,depth-1);
      GenerateRule(t.right,depth-1);
     END
  END
END
```

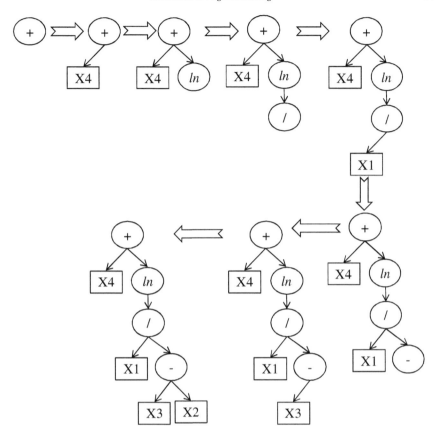

FIGURE 2.4: An example of creation of an S-expression in GP.

First, if the depth of the tree is one, a randomly chosen terminal is returned. Otherwise, it is randomly decided whether to select a function or a terminal. If a terminal is to be selected, it is randomly chosen from the list of terminals. In the case of a function, the tree gets either a left subtree or both subtrees depending on whether the function is a unary function or a binary function. This continues recursively.

An example of creating a tree of S-expression of the maximum depth five from the function and terminal sets of $\{+, -, *, /, \ln\}$ and $\{X1, X2, X3, X4\}$ is shown in Fig. 2.4. First, the addition function $+$ is chosen, which expects two arguments. In the next step, the terminal $X4$ is chosen randomly as the left argument of $+$ and ln as the right argument. Since ln is a unary function, $/$ is chosen randomly as its argument, which, in turn, requires two arguments. This continues until all the functions in the tree get their arguments.

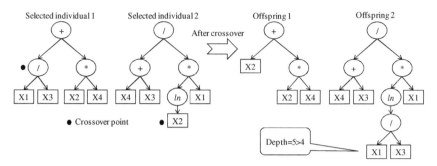

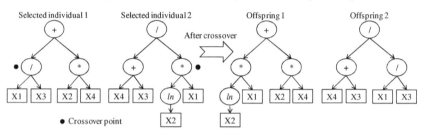

FIGURE 2.5: An example of crossover in GP.

2.1.8   Generation of Offspring through Crossover and Mutation

In Genetic Programming, the offspring are generated by applying crossover and mutation. For crossover, first, two individuals are selected from the population using a selection method, such as *fitness-proportionate selection* or *greedy over-selection* [Koza92]. Then the crossover is applied on the selected individuals with probability p_c, where p_c is the crossover probability. Next, two offspring are generated by exchanging the randomly chosen subtrees from the selected individuals. Usually, a subtree of the first individual is chosen randomly; then, in the second individual, a subtree is randomly chosen with the restriction that when it is exchanged with the subtree of the first individual will create two valid offspring of proper depth that is less than or equal to the maximum permitted depth. An example of generation of two offspring from two selected individuals with the maximum depth four is given in Fig. 2.5. In the first crossover, the first offspring has the depth of three, which is permitted but in the second offspring, the depth is five, which is greater than the maximum permitted depth. Therefore, this crossover will not be actually executed. In the second case, the depths of the two offspring are four and three, respectively, which are allowed; therefore, the crossover will be executed.

Mutation operates on a single individual. Usually, mutation is applied after the crossover operation and with a probability, which is very small compared to the crossover probability. There are various types of mutation in literature. Point mutation [Banzhaf *et al*.98] is one of them. In it, first, a node from the tree of the selected individual is randomly chosen. If the chosen node is a function, it is replaced with another function of the same type; if it is a terminal, it is replaced with another terminal. After the mutation operation, the depth of the tree remains the same; however, after the crossover operation, the depths of the trees may change.

2.1.9 Summary of GP Process

The following five basic elements are designed into a GP and enable it to handle a wide variety of application problems.

- Non-terminal nodes (functions in LISP S-expressions).

- Terminal symbols (atoms in LISP S-expressions; constants and variables that are function arguments).

- Fitness values.

- Parameters (crossover, probability of mutations, population sizes, etc.).

- Termination conditions.

It is also necessary to take care while constructing random tree structures for the initial population, because numbers of nodes and depth are not usually distributed uniformly throughout a tree. It is known that selection pressure does not work effectively when the production of trees is not uniform. The interested reader is directed to [Iba96c] for a more thorough discussion of these matters.

GP has been widely applied to many real-world problems, e.g., humanoid robots, music composition, bioinformatics and financial engineering. For instance, Fig. 2.6 and Fig. 2.7 show the successful cooperation of two HOAP-1 humanoid robots in the transportation task. In the case of object transportation involving two humanoid robots, mutual position shifts may occur due to the body swinging of the robots. Therefore, it is necessary to correct the position in real-time. Developing the position shift correction system requires a great deal of effort. Solution to the problem of learning the required behaviors is obtained by using GP-based method. Figs. 2.8, 2.9, and 2.10 show other demonstration systems developed in our laboratory. The readers should refer to [Ando and Iba05, Iba02, Iba *et al*.04] for more details.

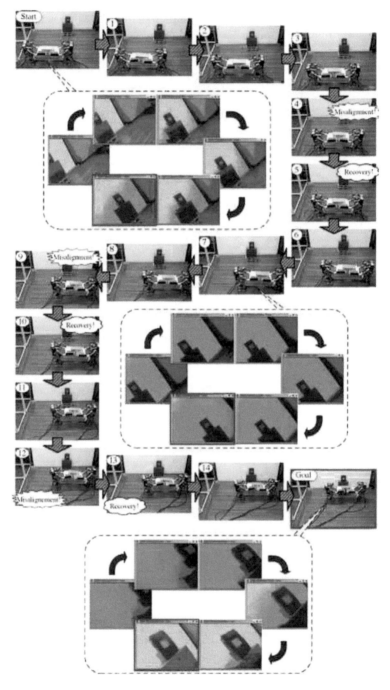

FIGURE 2.6: Cooperative transportation by evolutionary humanoid robots [Inoue *et al.*07].

FIGURE 2.7: Piano movers' task by humanoid robots [Iba *et al.*04].

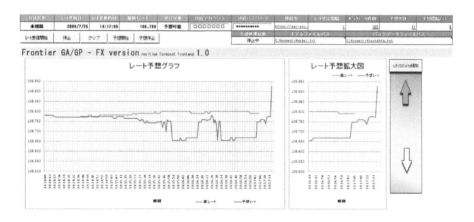

FIGURE 2.8: Frontier GA/GP-FX version: an overview of GP-based trading system courtesy of Medical Front, Inc.

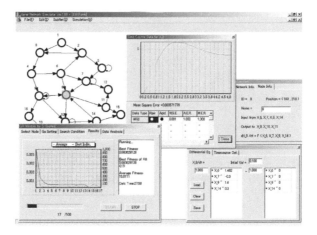

FIGURE 2.9: Interactive inference system for gene regulatory networks.

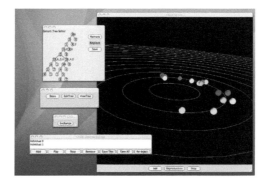

FIGURE 2.10: Music composition system by means of interactive GP.

2.2 LGPC System

This section examines the process of searching for a variety of types of programs using a GP simulator. This simulator is called LGPC (Linear GP in C), which has been developed in our laboratory for educational use.

LGPC is based on fast and efficient GP processes using a linear genome. It has been proposed that programs should be arranged in an array, so that the tree structures can be represented "implicitly", eliminating overhead such as the pointer operation. In this approach, called "Linear GP", rather than using a pointer with a program structure of a GP individual, the program is represented with a linear array [Keith and Martin94]. In other words, the program is expressed in a one-dimensional row of functions, constants, and variables. Linear GP allows not only a reduction in memory consumption from that used for tree structures in GP systems but also a reduction in the time required for pointer referencing. The purpose of Linear GP is to raise the speed and lower the memory load of programs by expressing their tree structures in arrays. The pointer is eliminated so that the time needed for referencing during assessments of GP individuals is avoided. The memory needed to store the pointer during execution is also saved. Details of this system are described in [Tokui and Iba99].

LGPC is available from the authors' homepage (see Appendix for more details). The basic function of the programs has been verified on several Windows operating systems, but a few problems may remain in some versions.

The next sections provide brief instructions on how to use the main features of this software.

2.2.1 Symbolic Regression (Regression Simulator)

Let us suppose that there is an unknown one-variable function f:

$$y = f(x) \tag{2.9}$$

and a set of n known values for input and output data,

$$\{(x_1, y_1), (x_2, y_2), \cdots, (x_n, y_n)\}, \tag{2.10}$$

where $y_i = f(x_i)$. The objective of identifying the function is to obtain an approximation $\overline{f}$ of the actual function f. This is an example of a system identification problem. It is also called a "symbolic regression" problem for approximation using a special functional symbol.

Let us try out a functional identification problem using the regression simulator (LGPC for Regression). Once the file has self-extracted, execute Regression.exe to observe a GP search in the context of a function identification problem (Fig. 2.11). After the data file for the function to be identified has

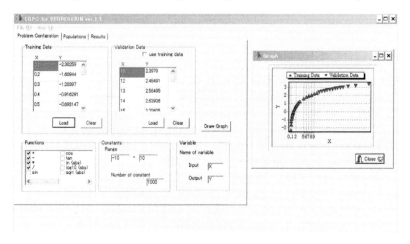

FIGURE 2.11: LGPC for Regression (1).

been read, it can be plotted by using the Draw Graph command. Click to select the desired non-terminal symbol. This window is also used to set the range for the random numbers. Fig. 2.12 shows the display for selecting the parameters. A wide variety of parameters used in the GP can be set here, e.g., population size, crossover ratio, mutation ratio, selection method, etc.

Select the Results window and click the Start button to initiate the search. The fitness value, program size and genotype (program) are displayed for each generation. The phenotype of the genes of the best individuals for the desired function (in other words, the values output by the best approximation functions) are shown together.

The fitness is equal to the mean squared error for this problem. The objective is to identify a function $y = f(x)$, and the fitness value is given by

$$\text{Fitness value} = \frac{1}{N} \sum_{i=1}^{N} (f(x_i) - \hat{f}(x_i))^2, \qquad (2.11)$$

where $\hat{f}(x)$ is the value of the function represented by the GP individual (genotype). The values on the x-axis for calculating the fitness are $x_1, x_2, \cdots, x_N$. Thus, low fitness values are desired.

Fig. 2.13 shows how this software handled a logarithmic curve in the Results window using log-training.txt file. In this search, the four arithmetic operations were employed as the non-terminal symbols for the GP. The other parameters were as follows: population size: 1000; maximum generations: 100; crossover ratio: 0.70; mutation ratio: 0.20, etc. The example search

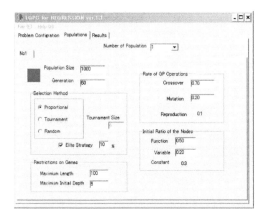

FIGURE 2.12: LGPC for Regression (2).

provided the following approximation:

$$Y = \cfrac{2X + 3.200}{X - \cfrac{X - (\frac{4.300}{X} - 2X) - (6.100 - X)}{2X + 3.200} - (X - (X + \frac{-0.400}{-5.400}) - 7.000 * 2.600) - 5.400}$$

$$Z = \cfrac{(X + \cfrac{-0.400}{(((-0.400 - (X + 2.000 - 7.000*2.600))*X) + 4.600)*(1.800 - 7.000*2.600 + X - 5.400)})}{1.800}$$

The above is a simplified form of the function displayed in the figure; it is broken into two parts to make it easier to display. The reader may be surprised at how long and complicated the obtained function is. Still, this approximation equation provides an excellent match to the data. Other validation data sets are also available with the program, and the user can also define data sets by him- or herself.

It is quite well known that programs (genotypes) tend to become increasingly long as the GP search process goes on. This is called "bloat" or "fluff", which means to become "structurally complex." This can be observed by watching the lower-left-hand plot (length of gene vs. generation). The reader can see that the graph slopes upward to the right. Bloat is one of the most persistent issues hindering the efficiency of GP searches. It would cause the following problems:

1. The large programs are difficult for people to understand.

2. The programs require much time and memory space to run.

3. Complicated programs tend to be inflexible and difficult to adapt to general cases, so that they are not very robust.

The following approaches are currently being used to control bloat:

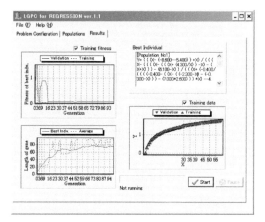

FIGURE 2.13: LGPC for Regression (3).

1. Set maxima for tree depth and size. Try to avoid creating tree structures exceeding these upper limits by means of crossover, mutation, etc. This is the easiest way to impose such controls, but the user needs to have a good understanding of the problem at hand, and needs heuristics in order to choose the appropriate settings for maxima.

2. Incorporate program size in the fitness value calculations, i.e., penalize large programs for being large. This is called "parsimony". More robust assessment standards using MDL (Minimum Description Length) and other parameters have been proposed (see next chapter for details).

3. Suppress the tree length by adjustments to genetic operators. For instance, Langdon proposed a homologous crossover or a size-fair crossover to control the tree growth [Langdon00].

When the Number of Populations in the Populations window is set to be a number greater than 1, it allows the user to perform simultaneous runs with multiple populations (however, parallel operation is not implemented). The operating parameters of the populations can be changed. For example, if one is using three populations, for the first, second and third populations respectively, one could set the size of population at 100, 200, 300, the crossover ratio at 0.5, and the mutation ratio at 0.2, 0.1, 0.0. Changing values for the parameters while executing these GP searches allows the user to compare parameter values for their effects.

2.2.2   Search for Food by an Artificial Ant (ANT Simulator)

The "Santa Fe trail" is a simple benchmark problem in artificial life. It is set up as follows. Food is placed in 89 locations in a 32×32 array of squares

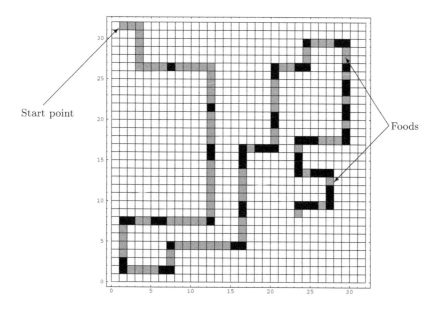

FIGURE 2.14: Santa Fe trail for ant problem.

Table 2.2: Function symbols in ant program.

Symbol	No. of Args.	Meaning
If_Food_Ahead	2	Execute 1st (2nd) argument if food is (not) straight ahead.
PROG2	2	Execute in order of 1st, 2nd arguments.
PROG3	3	Execute in order of 1st, 2nd, 3rd arguments.

(the gray squares in Fig. 2.14). The artificial ant searches for food within the limits of its available, limited energy. It can change directions and move ahead one step at a time. It cannot move diagonally or beyond a wall. It has a sensor that tells it if there is food in the square ahead of it. It uses one unit of energy every time it performs an action. Since the ant's energy is finite, it needs to search for food in the most efficient way possible. The optimal path is shown in deep gray in the diagram. The ant always begins from the upper left-corner. The non-terminal (function) symbol and terminal symbol are determined as shown in Tables 2.2 and 2.3 in order to produce a behavior program using the GP.

The fitness value of the ant is assessed in the following process.

Step1 Set Energy = 400, #food = 0.

Table 2.3: Terminal symbols in ant program.

Symbol	Meaning
RIGHT	Rotate 90 degrees right.
LEFT	Rotate 90 degrees left.
FORWARD	Move 1 step forward.

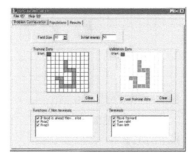

FIGURE 2.15: LGPC for ANT simulator (1).

Step2 The ant moves in accordance with the program. Once the FORWARD, LEFT or RIGHT terminal symbols have been assessed, set Energy = Energy − 1.

Step3 If there is food at the Ant's present location, set #food = #food + 1 and the food is deleted.

Step4 If Energy = 0, #food is returned as the (raw) fitness value.

Step5 Return to **Step2**.

In other words, the fitness assessment returns a value for the number of food supplies eaten.

Execute ANT.exe and observe the transition of the fitness values. The lower this value, the greater the success. Thus, if the fitness value is 0, it means that all of the food was eaten and a correct program was obtained. Fig. 2.15 presents the Problem Setting window for the program. This software permits the user to determine the size of the field and the locations of the food. Select the Results window and click the Start button to initiate the search (Fig. 2.16). The fitness value, program size, and optimal genotype (program) are displayed for each generation. The path followed by the best individual (how the food is eaten) is also displayed.

The following two kinds of data can be used for the simulation.

- Training Data
  These are the data used by the GP to calculate the fitness of an individual. They are also applied in searches and learning.

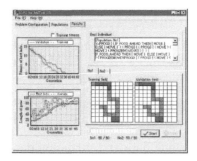

FIGURE 2.16: LGPC for ANT simulator (2).

- Validation Data

  These are the data used to evaluate the best (fittest) individual.

Validation data are employed to assess the robustness of solutions obtained. A program is said to be robust if it is resistant to noise and can be generalized to some extent. Let us consider what is important not only for GP but also for general learning processes. Consider, for example, a tutor teaching a student. If the drills are always the same or the homework does not have many problems, what happens? The student probably tries to learn the solutions by heart, and if a test has a problem that is not exactly like the drills (or unless the student makes some very lucky guesses), he or she will probably not do very well (and, no doubt, the tutor will be fired immediately). When a student has received full marks on the practice problems, but a bad grade on a test, this is called over-fitting or over-training. In order to gauge the student's ability (to judge the quality of the tutor), one must look at his or her test scores, not his or her homework (i.e., training) scores.

Well, then, what does the student need to do in order to get good test scores? A good tutor will probably prepare several drills for the student that show what to expect on tests, and train the student thoroughly in how to solve them. If the student is good, he or she will internalize the methods for solving practical problems in the course of doing the drills, and his or her test grades will show this. The most important element of this is to have the appropriate drills. When there is too large a gap between the drills and the test problems, study does not yield many results (and the student loses heart). Here, another term for "internalize the methods for solving practical problems" is "generalization", one of the most important indicators of the ability to learn. A characteristic of generalization of software is that a program can handle noise (errors in the data) in training examples. If a program has been over-trained, however, it overreacts to the presence of noise and cannot respond appropriately to validation data.

To summarize, the most important requirements for learning are as follows:

1. Training data and validation data are prepared in advance and are rea-

sonably similar to each other.

2. Training data are used for learning. Validation data are never used during the learning process.

3. The final evaluation is done with the validation data. It is assessed whether the program has acquired a sufficient level of generalization without over-fitting.

The usage of the training data and validation data is allowed on different maps for search problems for the ant. If the ant has been over-trained, bad results are obtained using the validation data.

There are generally noise and errors included, for a variety of reasons, in the data used for a learning process. Therefore, when the user simply conducts training (learning) until the fitness value (prediction error, etc.) has reached the minimum, the program has learned not only what was required in order to model the phenomenon of interest, it has also learned the errors in the particular set of data used for training. This prevents the new model from handling new data properly. In order for the model to be able to detect the nature of incoming data and obtain a robust solution without being influenced by the errors peculiar to the training data, training must be carried out very carefully. Let us consider what strategies will be most useful. (The remarks below are based on [Rowland03].)

As we mentioned above, training data and validation data are prepared in advance. The model displays good generality as long as the fitness value continues to rise. Learning should be stopped when the learning curves of both sets change from convergent to divergent (i.e., there is a transition of results). To put it another way, the model gets a better and better grasp of the details of the training data as the learning process goes on, so, up to a point, one can expect it to show steadily improving fitness. However, once it begins to learn the noise in the data, the fitness indicated after a run with the validation data generally begins to degrade. This is because the noise in the validation data differs from the noise in the training data. The two data sets to be modeled should share characteristics. Therefore, the region of the learning curve just before the point where it changes from convergent to divergent is the optimal point to stop learning. However, by this process, the validation data indirectly contribute to the learning process. If the user can predict the validation data, it is no longer possible to measure the robustness of the solution. Therefore, it is considered good practice to prepare a third validation data set (an independent data set) in order to make an assessment of a model that is completely independent of the training set. Hence three separate data sets should be used in order to ensure that the model is robust. The larger the data sets, the more robust the solution obtained. When the data sets are too small, the obtained model will not be very reliable, even if the procedure described below is followed, and the model may provide incorrect results if it is applied to an unknown data set.

There is a procedure called "cross validation", which is often employed when the data sets are small. Part of a data set is excluded and learning is performed with the remainder of the set. The excluded part is then used for the test. Next, this procedure is repeated with different portions excluded from the original data until all of the data has been excluded. For example, if a single trial uses 10% of the original data, this process can be followed ten times; learning is performed with the remaining 90% of the data each time, while each excluded 10% portion is used for the respective test. The mean value of the scores after the ten trials is an index of the robustness of the model.

Other procedures have been proposed:

- Bagging
  The learning process is repeated with randomly chosen portions from the training data.

- Boosting
  When the training data have resulted in false predictions (the learning failed), weighting of the elements of the training data is adjusted to allow learning to proceed correctly.

These have been applied in learning in some uses of GP. See [Iba99] for details.

2.2.3   Programming a Robot (Robot Simulator)

Let us attempt to evolve a program for controlling a robot using GP. The task here is to obtain a program that causes the robot to follow a wall (this could be useful for controlling a cleaning robot, since dust tends to gather against walls). The robot is capable of four motions: to move forward, move back, turn right, and turn left. It has a group of sensors that provide a 360-degree view. It is placed in an irregular room (i.e., an asymmetric room) and commanded to move about the periphery of the room, following the wall as accurately as possible.

This kind of program is easy to write in LISP [Koza91]. Let us begin with the adaptive learning necessary to solve this problem with the GP.

The robot is assumed to have 12 sonar sensors (s00 – s11) (see Fig. 2.17). The four motions permitted to the robot are straight ahead, straight back, turn left (+30 degrees), and turn right (−30 degrees). Since this GP carries out learning, the terminal nodes T are

$$T = \{S00, S01, \cdots, S11, \Re\}$$

where $S00, S01, \cdots, S11$ are the values output by the 12 distance sensors. $\Re$ is a random number variable (it is generated randomly when it is initially evaluated, i.e., it is generated once per terminal node). The non-terminal nodes F are

$$F = \{TR, TL, MF, MB, IFLTE, PROGN2\}$$

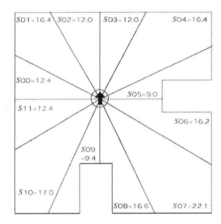

FIGURE 2.17: Robot sensors.

where TR and TL are functions for turning the robot 30 degrees right and left, respectively, and MF and MB are functions for moving the robot 1 foot forward or back, respectively. These functions do not accept arguments. It is assumed that execution of these functions takes 1 time unit, and that the sensor outputs are changed dynamically after execution. Two more functions are incorporated in order for the model to learn the appropriate control relationships. IFLTE (if-less-than-equal) takes four arguments and is interpreted as follows:

$$(\text{IFLTE } a \ b \ x \ y) \quad \Rightarrow \quad \text{if } a \leq b, \text{ then execute } x \text{ and return the result.}$$
$$\text{if } a > b, \text{ then execute } y \text{ and the result.}$$

PROGN2 takes 2 arguments, executes them in order and returns the value of the second.

Fig. 2.18 presents the Wall Following simulator by means of GP. The robot moves about the field displayed at left. Here, the green circles represent food (locations that the robot must pass through) and black locations represent obstacles. The check boxes at the bottom allow the user to set or delete these objects. When the robot passes over food, the box color is changed to purple. The path of the robot is written in blue when the robot is moving forward and red when it is moving in reverse. Use this simulator to create a program for the desired robot. The robot program will be executed according to a program displaying the GTYPE for a period of 400 time units. The fitness will be determined by how many of the round green circles (tiles) are crossed.

The path taken by the robot is shown by the test data and can be seen after the end of the run by clicking on the Test button. The test data can be revised using the Food, Block and Delete buttons mentioned above. The readers should check the generality of the evolved program by GP.

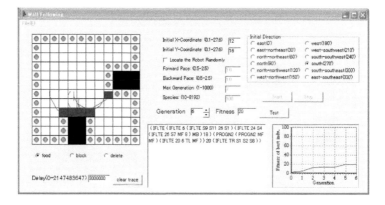

FIGURE 2.18: Wall Following by GP.

The generated program always has to cope with even small changes in the environment (such as shifts of the position of the chair in the room). As described in the previous section, this is what is meant by "robustness" of the program. It is not easy for a human to write this kind of program for a robot; in contrast, a GP can perform the necessary searches quite efficiently in order to write such a program.

Another similar example is Box Moving (SimBot) simulator (see Fig. 2.19). This has similar settings. The problem is for the robot to push the box to a wall, i.e., as if to put it away. It can be placed against any of the four walls. Since the box is treated as a solid body, however, it must be pushed in line with its center of gravity, or it will tend to spin while it is moving. The maximum number of initial locations permitted for the robot is four (in other words, the maximum number of training data groups is four), in order to prevent over-fitting. Try this simulator out in various environments.

2.2.4 Predicting Stock Prices: Time Series Prediction (TSP Simulator)

This simulator attempts to forecast time series. The objective here is to use the observed values for a function (time series data), i.e.,

$$x_1, x_2, x_3, \cdots, x_t \tag{2.12}$$

to obtain a function

$$x_t = f(x_{t-1}, x_{t-2}, x_{t-3}, x_{t-4}, \cdots, x_{t-M}), \tag{2.13}$$

which predicts the current value x_t from past data. Note that the time variable t is not included here as an argument. This is because we wish to identify a time series prediction function which is not dependent on the absolute time.

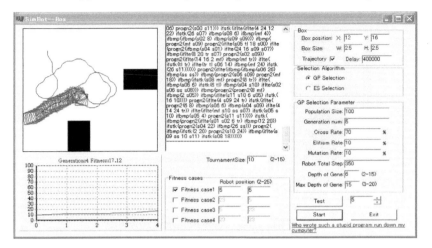

FIGURE 2.19: Box Moving by GP.

When learning with a GP, the user should define the window size, which designates how much past data will be used for the prediction (in other words, M in equation (2.13)). The fitness value is evaluated as the mean squared error between the prediction and the actually measured value. For example, if the window size is 5 and there are 50 training data elements $(x_1, x_2, \cdots, x_{50})$, then we obtain

$$\text{Fitness value} = \frac{1}{50} \sum_{t=6}^{50} \left[x_t - f(x_{t-1}, x_{t-2}, x_{t-3}, x_{t-4}, x_{t-5}) \right]^2 . \qquad (2.14)$$

The GP is trained so as to minimize the value of the above equation. Once training with the training data is complete, the term

$$\hat{f}(x_t, x_{t-1}, x_{t-2}, x_{t-3}, x_{t-4}) \qquad (2.15)$$

can be calculated for the predicted value x_{t+1} using the test data. $\hat{f}$ is the best individual evolved by GP that provides the lowest (best) fitness value.

Learning of time series data can be experienced in the GP simulator (LGPC for Time Series Prediction, see the Results window shown in Fig. 2.20). Refer to the on-line manual for details on how to operate the system. The user can define and load data by him- or herself. The following standard data sets have been provided for the sake of convenience (in the data folder created by the system).

- Log_training: Training data for logarithmic functions

- Log_validation: Validation data for logarithmic functions

- Sin_training: Training data for sinusoidal functions

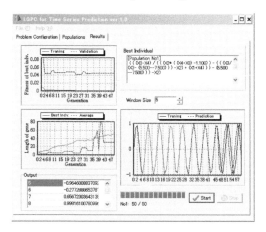

FIGURE 2.20: LGPC for TSP.

- Sin_validation: Validation data for sinusoidal functions

- Stock_training: Training data for Nikkei225 stock prices

- Stock_validation: Validation data for Nikkei225 stock prices

- Sunspot_training: Training data for the number of sunspots

- Sunspot_validation: Validation data for the number of sunspots

As mentioned earlier, the robustness of the obtained prediction program is assessed by its performance on the test data. After all, no matter how good a program is for predicting old stock prices from previous data (training data), if it has poor accuracy in predictions of future data (test data), it is useless.

Then, let us try a GP search for a function that predicts the Nikkei225 index. The Nikkei225 average is computed by the Nihon Keizai Shimbun-Sha, a well-known financial newspaper publishing firm in Japan (see Section 3.7.1.1 for the details). This package includes the Stock_training data set (the first 3,000 minutes, about half a month) and the Stock_validation data set (the remaining 30,177 minutes, about 5 months) for the tick data in Fig. 2.21 (first half of 1993). The degree of difference caused by normalizing the data can be used to assess the results. The application of GP and its extension to financial engineering will be described in Section 3.9.4.

2.2.5 Experiment in Pattern Recognition: Spiral Simulator

LGPC for Spiral simulator is used for the sake of solving pattern recognition problem (classification problem).

This section describes the experiment with a spiral problem [Angeline96]. This problem is to classify two data sets making up a two-dimensional spiral

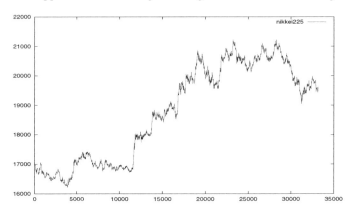

FIGURE 2.21: Nikkei225 data.

(Fig. 2.22). Each data set contains a total of 97 points.

$$\text{Class 1 (white points)} = \{(x_1^1, y_1^1), (x_2^1, y_2^1), \cdots, (x_{97}^1, y_{97}^1)\},$$
$$\text{Class 2 (black points)} = \{(x_1^2, y_1^2), (x_2^2, y_2^2), \cdots, (x_{97}^2, y_{97}^2)\}.$$

These data were produced with the following equations:

$$x_k^1 = 1 - x_k^2 = r_k \sin \alpha_k + 0.5,$$
$$y_k^1 = 1 - y_k^2 = r_k \cos \alpha_k + 0.5.$$

where

$$r_k = 0.4 \times \frac{105 - k}{104},$$
$$\alpha_k = \pi \times \frac{k - 1}{16}.$$

The data for this problem is in the file named as "spiral.txt." Once you have loaded the simulator, click Draw Graph command to get a two-dimensional display of the data to be recognized. Bear in mind that the two data sets contain the same numbers of Class 1 and 2 data points.

Functions box in the Problem Setting window can be used for the sake of selecting a variety of function arguments. The following terminal and non-terminal symbols are employed in this GP:

$$F = \{+, -, *, /, \text{IFLTE}, \sin, \cos\},$$
$$T = \{x, y, \Re\},$$

where x and y are the coordinates of the points to be classified. IFLTE is an if-less-than-equal function taking four arguments, i.e., (IFLTE a b c d) means that if a $\geq$ b, then c is executed; if a $<$ b, then d is executed. $\Re$ is a random number variable (it is generated randomly when it is initially evaluated).

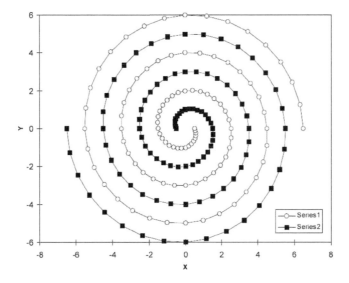

FIGURE 2.22: Spiral problem.

The goal is to obtain a function f by GP satisfying the following conditions:

$$f(x, y) > 0 \quad \Rightarrow \quad \text{white area},$$
$$f(x, y) < 0 \quad \Rightarrow \quad \text{black area}.$$

The fitness is designated by the fraction (%) of the 194 data points which are not correctly classified, i.e., the smaller this fraction, the better, and a score of 0 means the correct function has been obtained.

Press the Start button in the Results window to execute the GP. The fitness value, program size and genotype (program) of the best individual will be displayed for each succeeding generation (Fig. 2.23). The recognition capabilities by the best individual gene are plotted in a two-dimensional graph.

Add noise by editing one of the data files in spiral.txt. For example, change coordinate values for some points. Then, re-run the GP to see if you can obtain the desired classification function. This is to test the robustness of GP.

The data in xor.txt describes the XOR problem, which is often employed to test neural networks. This data cannot be classified by linear functions, i.e., no first-degree functions exist with any coefficients a and b satisfying the following condition:

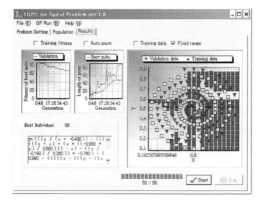

FIGURE 2.23: LGPC for Pattern Recognition.

$$ax + by > 0 \quad \Rightarrow \quad \text{black points,}$$
$$ax + by < 0 \quad \Rightarrow \quad \text{white points.}$$

These problems are called "linearly inseparable" and have long been recognized as unclassifiable by classical perceptrons. Conduct LGPC search to classify this data and observe how well the evolved function classifies the data.

2.2.6   Experiment in Noise: Image Compression Simulator

The Image Compression Simulator allows you to search for a function to compress images using a GP. This function approximates a given image with a function f fulfilling

$$c = f(x, y), \tag{2.16}$$

where coordinates x and y satisfy $-1 < x < 1, -1 < y < 1$ and c also satisfies $-1 < c < 1$. Fig. 2.24 shows how c is related to RGB (Red, Green, and Blue) . If $c = 0$, then (R, G, B) = (0, 255, 0). For instance, Fig. 2.25 shows the monitor display corresponding to the following function f:

$$f(x, y) = x^2 + y^2 - 0.95 \tag{2.17}$$

The goal of this problem is to find a function ($\hat{f}$) that approximates the given image. Once this function has been found, the approximation $\hat{f}$ can be used to transfer the image. Thus, this becomes a compressed form of the image, in a certain sense. There exists a correct solution f for the case in Fig. 2.25, but this is also a practical approach for handling images for which no f exists.

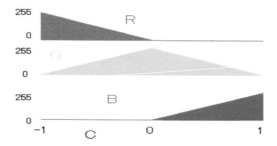

FIGURE 2.24: RGB encoding scheme.

FIGURE 2.25: Example display.

The fitness function is defined as the mean squared error for the function f. It is always positive, and the lower its value, the better.

Fig. 2.26 presents the results of execution of the GP. The left side of the Results window shows a plot of the fitness value and gene length of the best individual and the mean gene length in the population. The upper-right field shows the genetic code of the individual with the highest fitness, and the lower right displays represent the images of the training data and the test data written by the evolved functions.

The standard data are as follows:

- Circle: A small circle

- Circle2: A large circle

- Cup: A U curve

- Line1: A straight line (with gradual color changes)

- Line2: A straight line (with abrupt color changes)

- Line3: A straight line (with discontinuous color changes)

- S-curve: S-shaped curve

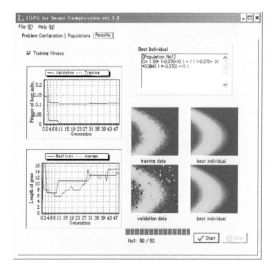

FIGURE 2.26: LGPC for Image Compression.

Click the Add Noise button in the training data and validation data to add noise to the image. The fraction of noise can be adjusted. As well as adding noise to the training data, the readers should experiment with the validation examples without noise. It is also interesting to change the size of the training and validation images.

2.2.7 Graphic Art

Regarding the optimization of a system, there are systems whose optimum performance can be expressed numerically as a fitness function, and systems that cannot be expressed numerically. For example, when a "favorite" outputting from human–machine interactive systems such as systems that create, process, and retrieve images and music, the desired output depends upon the subjectivity of the user, and it is generally difficult to assign a fitness function. Many such fields exist not only in the fields of art and aesthetics, but also in the fields of engineering and education.

One method that is used for such a system is Interactive Evolutionary Computation (IEC) [Takagi01]. IEC is a version of evolutionary computation, in which its fitness is evaluated directly through a human user. It is expected to be effectively applicable to developing a decision-support system. Traditionally, human evaluation is analytically modeled and incorporated into the system. On the other hand, the human him- or herself is included in the framework of IEC-based design process and he or she directs the evolutionary process according to his or her criterion, as a result of which an optimal design is evolvable.

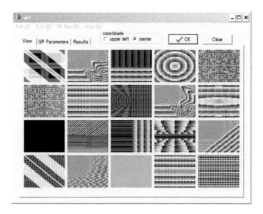

FIGURE 2.27: LGPC for Art.

The procedure for using IEC consists of the following steps:

Step1 Initialize a gene group consisting of N individuals using random information.

Step2 Generate and display the type of expression of each individual from genes.

Step3 Select the desired individual from the displayed individuals, based on the subjectivity of the user.

Step4 Create genes for N individuals using genetic manipulation, and create a new gene group, taking the genes of the selected individual as the parent genes.

Step5 Return to **Step**2.

With the exception of the selection part, the above is exactly the same as the procedure for a general GP (Fig. 2.3).

SBART [Unemi99] is an IEC-based software package for the purpose of creating two-dimensional graphic art with a GP. A three-dimensional vector equation including variables x and y is used to draw images in SBART. The x- and y-coordinates of every point (pixel) in an image are substituted into an equation and the value returned is transformed into image information. This process produces psychedelic, beautiful two-dimensional images.

Fig. 2.27 shows the LGPC for Art, which is a simple implementation of SBART. LGPC for Art is an interactive GP tool for creating abstract images (or wallpaper designs). It has the following basic procedure:

Step1 Click Clear if you do not like any of the pictures in the 20 windows. This re-initializes all of the windows.

Step2 Click one of the pictures to select it if you like it. The frame will turn red. You can choose as many as you want.

Step3 Press the OK button to display the population of the next generation of genes produced from the image you selected.

Step4 Repeat from **Step1** to **Step3**.

Any image you like can be stored (with the Gene_Save command) and any stored image can be loaded (with the Gene_Load command) to replace the currently displayed image. The 20 original windows loaded with the View tab are numbered (#1 is at the upper left, and numbering proceeds left to right, then up to down).

Play with this process to "cultivate" new images. One of the authors has actually used this system to create the cover design for a book. The design process, which usually requires the professional designer several days to prepare a camera-ready image, was completed in several hours.

Chapter 3

Numerical Approach to Genetic Programming

3.1 Introduction

This chapter introduces a new approach to Genetic Programming (GP), based on a numerical, i.e., GMDH(Group Method of Data Handling)-based, technique, which integrates a GP-based adaptive search of tree structures, and a local parameter tuning mechanism employing statistical search (i.e., a system identification technique). In traditional GP, recombination can cause frequent disruption of building blocks, or mutation can cause abrupt changes in the semantics. To overcome these difficulties, we supplement traditional GP with a local hill-climbing search, using a parameter tuning procedure. More precisely, we integrate the structural search of traditional GP with a multiple regression analysis method and" establish our adaptive program called "STROGANOFF" (i.e., STructured Representation On Genetic Algorithms for NOnlinear Function Fitting). The fitness evaluation is based on a "Minimum Description Length " (MDL) criterion, which effectively controls the tree growth in GP. We demonstrate its effectiveness by solving several system identification (numerical) problems and compare the performance of STROGANOFF with traditional GP and another standard technique (i.e., "radial basis functions"). The effectiveness of this numerical approach to GP is demonstrated by successful application to computational finance.

3.2 Background

The target problem we solve is "system identification." Attempts have been made to apply traditional GP to the system identification problems, but difficulties have arisen due to the fact that GP recombination can cause frequent disruption of building blocks[1], or that mutation can cause abrupt changes in the semantics. We convert a symbolic (discrete) search problem into a numeric (continuous) search space problem (and vice versa).

3.2.1   System Identification Problems

A system identification problem is defined in the following way. Assume that a single valued output y, of an unknown system, behaves as a function of m input values, i.e.,

$$y = f(x_1, x_2, \cdots, x_m). \tag{3.1}$$

Given N observations of these input–output data pairs, i.e.,

INPUT				OUTPUT
x_{11}	x_{12}	$\cdots$	x_{1m}	y_1
x_{21}	x_{22}	$\cdots$	x_{2m}	y_2
		$\cdots$		$\cdots$
x_{N1}	x_{N2}	$\cdots$	x_{Nm}	y_N

the system identification task is to approximate the function f with an approximate function $\overline{f}$ called the "complete form".

There is a wide range of applications for system identification. An example of system identification is time series prediction, i.e., predicting future values of a variable from its previous values (see Fig. 3.7 for instance). Expressed in system identification terms, the output $x(t)$ at time t is to be predicted from its values at earlier times $(x(t-1), x(t-2), \cdots)$, i.e.,

$$x(t) = f(x(t-1), x(t-2), x(t-3), x(t-4), \cdots). \tag{3.2}$$

Another example is a type of pattern recognition (or classification) problem, in which the task is to classify objects having m features $x_1, \cdots, x_m$ into one of two possible classes, i.e., "C" and "not C." If an object belongs to class C, it is said to be a positive example of that class, otherwise it is a negative example. In system identification terms, the task is to find a (binary) function f of the m features of objects such that

$$y = f(x_1, x_2, \cdots, x_m) = \begin{cases} 0, & \text{negative example} \\ 1, & \text{positive example.} \end{cases} \tag{3.3}$$

The output y is 1 if the object is a positive example (i.e., belongs to class C), and y is 0 if the object is a negative example.

Most system identification techniques are based on parameter and function estimates. Unfortunately these earlier approaches suffered from combinatorial explosion as the number of training data, parameters, and constrained

---

[1]In this section, a building block (i.e., schema) for GP is defined as a subtree which is a part of a solution tree.

assumptions increased. One of these approaches was a heuristic algorithm called GMDH (Group Method of Data Handling) [Ivakhnenko71]. It too had its weaknesses, due to its heuristic nature, e.g. it suffered from local optima problems, which limited its application [Tenorio *et al*.90]. However, this chapter shows that the weakness of the GMDH approach can be largely overcome by wedding it to a (structured) GP-based approach.

3.2.2 Difficulties with Traditional GP

GP searches for desired tree structures by applying genetic operators such as crossover and mutation. However, standard GP is faced with the following difficulties in terms of efficiency.

1. A lack of tools to guide the effective use of genetic operators.

2. Representational problems in designing node variables.

3. Performance evaluation of tree structures.

First, traditional GP blindly combines subtrees, by applying crossover operations. This blind replacement, in general, can often disrupt beneficial building blocks in tree structures. Randomly chosen crossover points ignore the semantics of the parent trees. For instance, in order to construct the Pythagoras relation (i.e., $a^2 + b^2 = c^2$) from two parent trees (i.e., $c^2 = 3$ and $(a - c) \times (a^2 + b^2)$), only one pair of crossover points is valid (see Fig. 3.1(b)). Thus crossover operations seem almost hopeless as a means to construct higher-order building blocks. [Koza90, ch.4.10.2] used a constrained crossover operator when applied to neural network learning, in which two types of node (weight and threshold functions) always appeared alternately in trees to represent feasible neurons. A constrained crossover operator was applied so that it would preserve the order constraint. It worked well for neural network learning but its applicability is limited. [Schaffer and Morishima87] discussed adaptive crossover operations for usual string-based GA (Genetic Algorithm). Although the quality of adaptation is also desirable for effective search in GP, it is difficult to implement within the usual GP framework.

Second, choosing a good representation (i.e., designing the terminal set $\{T\}$ and functional set $\{F\}$) is essential for GP search. Recombination operations (such as swapping subtrees or nodes) often cause radical changes in the semantics of the trees. For instance, the mutation of the root node in Fig. 3.1(a) converts a Boolean function to a totally different function, i.e., from false $((x \wedge y) \wedge (\overline{x} \vee \overline{y}) \equiv 0)$ to true $((x \wedge y) \vee (\overline{x} \vee \overline{y}) \equiv 1)$. We call this phenomenon "semantic disruption", which is due to the "context-sensitive" representation of GP trees. As a result, useful building blocks may not be able to contribute to higher fitness values of the whole tree, and the accumulation of schemata may be disturbed. To avoid this, [Koza94] proposed a strategy called ADF (Automatic Defining Function) for maintenance of useful building blocks.

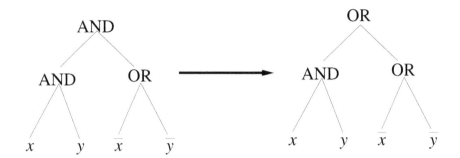

(a) Gmutation (node label change or subtree substitution).

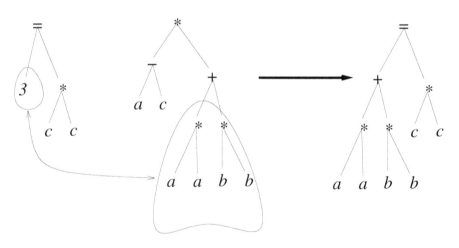

(b) Gcrossover (subtree swapping).

FIGURE 3.1: Genetic operators for GP [Iba *et al.*96a, Fig. 1].

Third, the fitness definitions used in traditional GP do not include evaluations of the tree descriptions. Therefore without the necessary control mechanisms, trees may grow exponentially large or become so small that they degrade search efficiency. Usually the maximum depth of trees is set as a user-defined parameter in order to control tree sizes, but an appropriate depth is not always known beforehand.

3.2.3 Numerical Approach to GP

To overcome the above difficulties, this chapter introduces a new GP-based approach to solving system identification problems, by establishing an adaptive system we call "STROGANOFF". STROGANOFF integrates a multiple regression analysis method and a GP-based search strategy. Its fitness definition is based upon an MDL criterion. The theoretical basis for this work is derived from a system identification technique due to Ivakhnenko [Ivakhnenko71].

The advantages of STROGANOFF are summarized as follows:

1. GP search is effectively supplemented with the tuning of node coefficients by multiple regression.

2. Analogue (i.e., polynomial) expressions complemented the digital (symbolic) semantics. Therefore, the representational problem of standard GP does not arise for STROGANOFF.

3. MDL-based fitness evaluation works well for tree structures in STROGANOFF, which controls GP-based tree search.

The effectiveness of this numerical approach is demonstrated both by successful application to numeric and symbolic problems, and by comparing STROGANOFF's performance with a traditional GP system, applied to the same problems.

3.3 Principles of STROGANOFF

STROGANOFF consists of two adaptive processes; a) The evolution of structured representations, using a traditional Genetic Algorithm, b) The fitting of parameters of the nodes with a multiple regression analysis. The latter part is called a GMDH process, which will be explained in the next section.

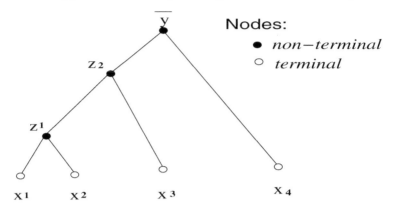

FIGURE 3.2: GMDH network.

3.3.1   GMDH: Group Method of Data Handling

GMDH is a multiple-variable analysis method which is used to solve system identification problems [Ivakhnenko71]. This method constructs a feedforward network (as shown in Fig. 3.2) as it tries to estimate the output function $\bar{y}$. The node transfer functions (i.e., the Gs in Fig. 3.3) are quadratic polynomials of the two input variables (e.g., $G(z_1, z_2) = a_0 + a_1 z_1 + a_2 z_2 + a_3 z_1 z_2 + a_4 z_1^2 + a_5 z_2^2$) whose parameters a_i are obtained using regression techniques. GMDH uses the following algorithm to derive the "complete form" $\bar{y}$.

Step1 Let the input variables be $x_1, x_2, \cdots, x_m$ and the output variable be y. Initialize a set labelled VAR using the input variables, i.e., $VAR := \{x_1, x_2, \cdots, x_m\}$.

Step2 Select any two elements z_1 and z_2 from the set VAR. Form an expression $G_{z1,z2}$ which approximates the output y (in terms of z_1 and z_2) with least error using multiple regression techniques. Regard this function as a new variable z,

$$z = G_{z1,z2}(z_1, z_2). \tag{3.4}$$

Step3 If z approximates y better than some criterion, set the "complete form" (i.e., $\bar{y}$) as z and terminate.

Step4 Else $VAR := VAR \cup \{z\}$. Go to **Step2**.

The "complete form" (i.e., $\bar{y}$) given by the GMDH algorithm can be represented in the form of a tree. For example, consider the GMDH network of Fig. 3.2, where the "complete form" $\bar{y}$ is given as follows:

$$z_1 = G_{x1,x2}(x_1, x_2), \tag{3.5}$$

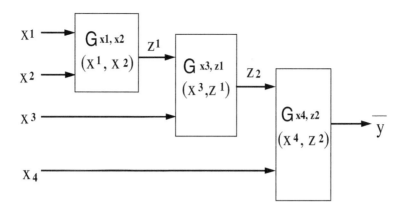

FIGURE 3.3: Equivalent binary tree.

$$z_2 = G_{x3,z1}(x_3, z_1), \tag{3.6}$$

and

$$\bar{y} = G_{x4,z2}(x_4, z_2). \tag{3.7}$$

where equations (3.5) and (3.6) are subexpressions given in **Step2** of the GMDH algorithm. By interpreting these subexpressions as non-terminal symbols and the original input variables $\{x_1, x_2, x_3, x_4\}$ as terminal symbols, we get the tree shown in Fig. 3.3.

An example of G expression takes the following quadratic form.

$$G_{z1,z2}(z_1, z_2) = a_0 + a_1 z_1 + a_2 z_2 + a_3 z_1 z_2 + a_4 z_1^2 + a_5 z_2^2. \tag{3.8}$$

If z_1 and z_2 are equal (i.e., $z_1 = z_2 = z$), G is reduced to

$$G_z(z) = a_0' + a_1' z + a_2' z^2. \tag{3.9}$$

The coefficients a_i are calculated using a least mean square method, which will be explained in Section 3.3.3 in details.

3.3.2 STROGANOFF Algorithm

In summary, STROGANOFF algorithm is described below:

Step1 Initialize a population of tree expressions.

Step2 Evaluate each expression in the population so as to derive the MDL-based fitness (Section 3.3.6, equation (3.32)).

Step3 Create new expressions (children) by mating current expressions. With a given probability, apply mutation and crossover (Figs. 3.1 and 3.6) to generate the child tree expressions (Sections 3.3.4, 3.3.5 and 3.3.8).

Step4 Replace the members of the population with the child trees.

Step5 Execute the GMDH process, so as to compute the coefficients of the intermediate nodes of the child trees (Section 3.3.3, equation (3.17)).

Step6 If the termination criterion is satisfied, then halt; else go to **Step2**.

In **Step5**, the coefficients of the child trees are recalculated using the GMDH process. However, this recalculation is performed only on intermediate nodes, upon whose descendants crossover or mutation operators were applied. Therefore, the computational cost of **Step5** is expected to be reduced as the generations proceed. As can be seen, **Steps**1~4 and **Step6** follow traditional GP, whereas **Step5** is the new local hill-climbing procedure, which will be discussed in Section 3.9.

3.3.3 GMDH Process in STROGANOFF

STROGANOFF constructs a feedforward network, as it estimates the output function $\bar{f}$. The node transfer functions are simple (e.g., quadratic) polynomials of the two input variables, whose parameters are obtained using regression techniques.

An example of a binary tree generated by STROGANOFF is shown in Fig. 3.4. For instance, the upper-left parent tree (P_1) can be written as an S-expression in LISP,

(NODE1
    (NODE2
        (NODE3 (x_1) (x_2))
        (x_3)
    (x_4)))

where x_1, x_2, x_3, x_4 are the input variables. Intermediate nodes represent simple polynomial relationships between two descendant (lower) nodes. This tree expresses a "complete form" $\bar{y}$ given by the GMDH process as follows:

1. Select two variables x_1 and x_2 and form an expression $G_{x1,x2}$ which approximates the output y (in terms of x_1 and x_2) with the least error using the multiple regression technique. Regard this function as a new variable z_1 (i.e., the new intermediate node NODE3),

$$z_1 = G_{x1,x2}(x_1, x_2). \tag{3.10}$$

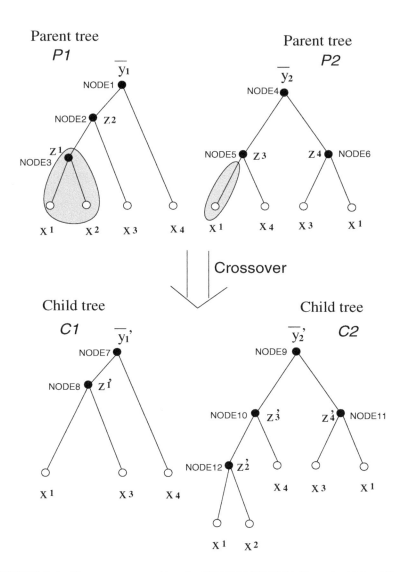

FIGURE 3.4: Crossover operation in STROGANOFF [Iba *et al.*96a, Fig. 2].

2. Select two variables z_1 and x_3 and form an approximating expression $G_{z1,x3}$ in the same way. Regard this function as a new variable z_2 (i.e., the new intermediate node NODE2),

$$z_2 = G_{z1,x3}(z_1, x_3). \tag{3.11}$$

3. Select two variables z_2 and x_4 and form an approximating expression $G_{z2,x4}$. Regard this function as a "complete form" $\overline{y}$, (i.e., the root node NODE1),

$$\overline{y} = G_{z2,x4}(z_2, x_4). \tag{3.12}$$

For the sake of simplicity, this section assumes quadratic expressions for the intermediate nodes. Thus, each node records the information derived by the following equations:

$$NODE3 \ : \ z_1 = a_0 + a_1 x_1 + a_2 x_2 + a_3 x_1 x_2 + a_4 x_1^2 + a_5 x_2^2, \tag{3.13}$$

$$NODE2 \ : \ z_2 = b_0 + b_1 z_1 + b_2 x_3 + b_3 z_1 x_3 + b_4 z_1^2 + b_5 x_3^2, \tag{3.14}$$

$$NODE1 \ : \ \overline{y_1} = c_0 + c_1 z_2 + c_2 x_4 + c_3 z_2 x_4 + c_4 z_2^2 + c_5 x_4^2, \tag{3.15}$$

where z_1 and z_2 are intermediate variables, and $\overline{y_1}$ is an approximation of the output, i.e., the complete form. These equations are called "subexpressions." All coefficients $(a_0, a_1, \cdots, c_5)$ are derived from multiple regression analysis using a given set of observations. For instance, the coefficients a_i in equation (3.13) are calculated using the following least mean square method. Suppose that N data triples (x_1, x_2, y) are supplied from observation, e.g.,

$$x_{11} \ \ x_{21} \ \ y_1$$
$$x_{12} \ \ x_{22} \ \ y_2$$
$$\cdots$$
$$x_{1N} \ \ x_{2N} \ \ y_N$$

From these triples, an X matrix is constructed,

$$X = \begin{pmatrix} 1 & x_{11} & x_{21} & x_{11}x_{21} & x_{11}^2 & x_{21}^2 \\ 1 & x_{12} & x_{22} & x_{12}x_{22} & x_{12}^2 & x_{22}^2 \\ & & & \cdots & & \\ 1 & x_{1N} & x_{2N} & x_{1N}x_{2N} & x_{1N}^2 & x_{2N}^2 \end{pmatrix} \tag{3.16}$$

which is used to define a coefficient vector $\mathbf{a}$, given by

$$\mathbf{a} = (X'X)^{-1}X'\mathbf{y} \tag{3.17}$$

where

$$\mathbf{a} = (a_0, a_1, a_2, a_3, a_4, a_5)' \tag{3.18}$$

and

$$\mathbf{y} = (y_1, y_2, \cdots, y_N)', \tag{3.19}$$

where X' is the transposed matrix of X. All coefficients a_i are so calculated that the output variable z_1 approximates the desired output y. The other coefficients are derived in the same way.

Note that all node coefficients are derived locally. For instance, consider b_i's of NODE2. When applying the multiple regression analysis to the equation (3.14), these b_i's are calculated from the values of z_1 and x_3 (i.e., the two lower nodes), not from x_4 or $\overline{y_1}$ (i.e., the upper node). Therefore, the GMDH process in STROGANOFF can be regarded as a local hill-climbing search, in the sense that the coefficients of a node are dependent only on its two descendant (lower) nodes.

If the determinant of $X'X$ is zero (i.e., $X'X$ is a singular matrix), $X'X^{-1}$ should be replaced by the Moore-Penrose generalized inverse matrix $X'X^+$. Thus, we get the following equation,

$$\mathbf{a} = (X'X)^+ X' y. \tag{3.20}$$

A Moore-Penrose generalized inverse matrix is a coefficient matrix which gives a minimal-norm solution to a least square problem. [Spiegel75] and [Press *et al.*88] should be referred to for the details of this process and its theoretical explanation.

3.3.4 Crossover in STROGANOFF

We now consider the recombination of binary trees in STROGANOFF. Suppose two parent trees P_1 and P_2 are selected for recombination (Fig. 3.4). Besides the above equations, internal nodes record polynomial relationships as listed below:

$$NODE5 : z_3 = d_0 + d_1 x_1 + d_2 x_4 + d_3 x_1 x_4 + d_4 x_1^2 + d_5 x_4^2, \tag{3.21}$$

$$NODE6 : z_4 = e_0 + e_1 x_3 + e_2 x_1 + e_3 x_3 x_1 + e_4 x_3^2 + e_5 x_1^2, \tag{3.22}$$

$$NODE4 : \overline{y_2} = f_0 + f_1 z_3 + f_2 z_4 + f_3 z_3 z_4 + f_4 z_3^2 + f_5 z_4^2. \tag{3.23}$$

Suppose z_1 in P_1 and x_1 in P_2 (shaded portions in Fig. 3.4) are selected as crossover points in the respective parent trees. This gives rise to the two child trees C_1 and C_2 (lower part of Fig. 3.4). The internal nodes represent the following relations:

$$NODE8 \; : \; z_1' = a_0' + a_1'x_1 + a_2'x_3 + a_3'x_1x_3 + a_4'x_1^2 + a_5'x_3^2, \qquad (3.24)$$

$$NODE7 \; : \; \overline{y_1}' = b_0' + b_1'z_1' + b_2'x_4 + b_3'z_1'x_4 + b_4'z_1'^2 + b_5'x_4^2, \qquad (3.25)$$

$$NODE12 \; : \; z_2' = c_0' + c_1'x_1 + c_2'x_2 + c_3'x_1x_2 + c_4'x_1^2 + c_5'x_2^2, \qquad (3.26)$$

$$NODE10 \; : \; z_3' = d_0' + d_1'z_2' + d_2'x_4 + d_3'z_2'x_4 + d_4'z_2'^2 + d_5'x_4^2, \qquad (3.27)$$

$$NODE11 \; : \; z_4' = e_0' + e_1'x_3 + e_2'x_1 + e_3'x_3x_1 + e_4'x_3^2 + e_5'x_1^2, \qquad (3.28)$$

$$NODE9 \; : \; \overline{y_2}' = f_0' + f_1'z_3' + f_2'z_4' + f_3'z_3'z_4' + f_4'z_3^{2} + f_5'z_4'^2. \qquad (3.29)$$

Since these expressions are derived from multiple regression analysis, we have the following equations:

$$z_2' = z_1, \qquad (3.30)$$
$$z_4' = z_4. \qquad (3.31)$$

Thus, when applying crossover operations, we need only derive polynomial relations for $z_1', z_3', \overline{y_1}', \overline{y_2}'$. In other words, recalculation of the node coefficients for the replaced subtree (z_2') and non-replaced subtree (z_4') is not required, which reduces much of the computational cost in STROGANOFF.

3.3.5   Mutation in STROGANOFF

When applying mutation operations, we consider the following cases:

1. A terminal node (i.e., an input variable) is mutated to another terminal node (i.e., another input variable).

2. A terminal node (i.e., an input variable) is mutated to a non-terminal node (i.e., a subexpression).

3. A non-terminal node (i.e., a subexpression) is mutated to a terminal node (i.e., an input variable).

4. A non-terminal node (i.e., a subexpression) is mutated to another non-terminal node (i.e., another subexpression).

3.3.6   Fitness Evaluation in STROGANOFF

STROGANOFF uses an MDL-based fitness function for evaluating the tree structures. This fitness definition involves a trade-off between certain structural details of the tree, and its fitting (or classification) errors.

$$\text{MDL fitness} = (Tree\_Coding\_Length) + (Exception\_Coding\_Length).$$
$$(3.32)$$

MDL fitness definition for our binary tree is defined as follows [Tenorio *et al.*90]:

$$Tree\_Coding\_Length = 0.5k \log N, \qquad (3.33)$$

$$Exception\_Coding\_Length = 0.5N \log S_N^2, \qquad (3.34)$$

where N is the number of input–output data pairs, S_N^2 is the mean square error, i.e.,

$$S_N^2 = \frac{1}{N} \sum_{i=1}^{N} | \overline{y_i} - y_i |^2, \qquad (3.35)$$

and k is the number of parameters of the tree, e.g., the k-value for the tree P_1 in Fig. 3.4 is $6 + 6 + 6 = 18$ because each internal node has six parameters $(a_0, \cdots, a_5$ for $NODE3$ etc.).

An example of this MDL calculation is given in Section 3.4.1.

3.3.7   Overall Flow of STROGANOFF

The STROGANOFF algorithm is described below:

Input: t_{max}, I, Pop_size
Output: x, the best individual ever found

1 $t \leftarrow 0$;
   {I is a set of input variables (see equation (3.1)).
   $NODE\_2$ is a non-terminal node of 2-arity.}
2 $P(t) \leftarrow$ initialize($Pop\_size, I, \{NODE\_2\}$);
3 $F(t) \leftarrow$ evaluate($P(t), Pop\_size$);
4 $x \leftarrow a_j(t)$ **and** $Best\_so\_far \leftarrow MDL(a_j(t))$, where $MDL(a_j(t)) = \min(F(t))$;
   {the main loop of selection, recombination, mutation.}
5 **while** $(\iota(P(t), F(t), t_{max}) \neq$ true) **do**
6   **for** $i \leftarrow 1$ **to** $\frac{Pop\_size}{2}$ **do**
                {select parent candidates according to the MDL values.}
                $Parent_1 \leftarrow select(P(t), F(t), Pop\_size)$;
                $Parent_2 \leftarrow select(P(t), F(t), Pop\_size)$;
                {apply GP crossover operation, i.e., swapping subtrees (Fig. 3.4).}

$$a'_{2i-1}(t), a'_{2i}(t) \leftarrow \text{GP\_recombine}(Parent_1, Parent_2);$$
{apply GP mutation operation,
i.e., changing a node label and deleting/inserting a subtree.}
$$a''_{2i}(t) \leftarrow \text{GP\_mutate}(a'_{2i}(t));$$
$$a''_{2i-1}(t) \leftarrow \text{GP\_mutate}(a'_{2i-1}(t));$$
    od

7    $P''(t) \leftarrow (a''_1(t), \cdots, a''_{Pop\_size}(t));$

8    $F(t) \leftarrow \text{evaluate}(P''(t), Pop\_size);$

9    $tmp \leftarrow a''_k(t),$ where $MDL(a''_k(t)) = \min(F(t));$

10   **if** $(Best\_so\_far > MDL(a''_k(t)))$
        then $x \leftarrow tmp$ **and** $Best\_so\_far \leftarrow MDL(a''_k(t));$

11   $P(t+1) \leftarrow P''(t);$

12   $t \leftarrow t+1;$
  od
  return $(x);$

{terminate if more than t_{max} generations are over.}
1 $\iota(P(t), F(t), t_{max}):$
2   **if** $(t > t_{max})$
        then return *true*;
        else return *false*;

{initialize the population randomly.}
1 initialize(Pop_size, T, F):
2   **for** $i \leftarrow 1$ **to** Pop_size **do**
        generate a tree a_i randomly,
          where the terminal and non-terminal sets are T and F.
    od
    return $(a_1, \cdots, a_{Pop\_size});$

{evaluate of a population of size Pop_size.}
1 evaluate($P(t), Pop\_size$):
2   **for** $i \leftarrow 1$ **to** Pop_size **do**
        {calculate equation (3.35).}
        GMDH_Process(a_i);
        $S_N^2(a_i) \leftarrow$ the mean square error of a_i;
        {calculate equations (3.32),(3.33) and (3.34).}
        $MDL(a_i) \leftarrow Tree\_Coding\_Length(a_i)$
               $+ Exception\_Coding\_Length(a_i);$
    od
    return $(MDL(a_1), \cdots, MDL(a_{Pop\_size}));$

{execute the GMDH process.}
1 GMDH_Process(a):
2   nd ← the root node of a;
3   **if** (nd is a terminal node)
                 then return;
            {if the node coefficients of nd are already derived, then return.}
4   **if** ($Coeff(nd) \neq NULL$)
                 then return;
5   nl ← left_child(nd);
6   nr ← right_child(nd);
7   GMDH_Process(nl);
8   GMDH_Process(nr);
9          $Coeff(nd)$ ← Mult_Reg(nl, nr);
            return;

{execute the multiple regression analysis.}
1 Mult_Reg($n1, n2$):
            Assume $n1$ is the first variable and $n2$ is the second variable.
            For instance, x_1 ← $n1, x_2$ ← $n2$ for equation (3.13)
            Derive and return the fitting coefficients, i.e., equation (3.18)
            return;

In the GMDH_Process called by the evaluate routine, the coefficients of the child trees are recalculated using the multiple regressions. However, this recalculation is performed only on intermediate nodes, upon whose descendants crossover or mutation operators were applied (see the fourth lines in GMDH_Process). Therefore, the computational burden of the GMDH process is expected to be reduced as the generations proceed. As can be seen, lines from 6 to 7 in the STROGANOFF algorithm follow traditional GP, whereas GMDH_Process is the new local hill-climbing procedure, which will be discussed in Section 3.9.3.

3.3.8 Recombination Guidance in STROGANOFF

Multiple regressions in STROGANOFF tune the node coefficients so as to guide GP recombination effectively with MDL values. By this mechanism, STROGANOFF can avoid the disruption problem caused by the traditional GP crossover or mutation (Fig. 3.4). This section explains the recombination guidance of STROGANOFF.

Fig. 3.5 illustrates an exemplar STROGANOFF tree for the time series prediction (see Section 3.4.1 for details), in which the error of fitness ratios (i.e., mean square error, MSE) and MDL values are shown for all subtrees. As can be seen from the figure, the MSE values monotonically decrease towards the root node in a given tree. Thus the root node has the lowest (i.e., best) MSE

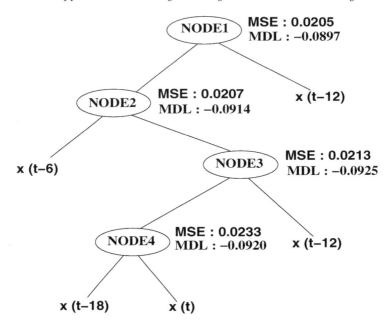

FIGURE 3.5: An exemplar STROGANOFF tree [Iba *et al.*96a, Fig. 3].

value. However, the MDL values do not monotonically change. The subtree whose MDL value is lowest is expected to give the best performance of all subtrees. Therefore, it can work as a building-block for crossover operations.

We realize a type of adaptive recombination based on MDL values. For this purpose, in applying crossover or mutation operators, we follow the rules described below:

1. Apply a mutation operator to a subtree whose MDL value is larger.

2. Apply a crossover operator to a subtree whose MDL value is larger, and get a subtree whose MDL value is smaller from another parent.

When the second operator is applied to two parents P_1 and P_2, execute the following steps (see Fig. 3.6).

1. Let $W1$ and $W2$ be the subtrees with the largest MDL values of $P1$ and $P2$.

2. Let $B1$ and $B2$ be the subtrees with the smallest MDL values of $P1$ and $P2$.

3. A new child $C1$ is a copy of $P1$, in which $W1$ is replaced by $B2$.

4. A new child $C2$ is a copy of $P2$, in which $W2$ is replaced by $B1$.

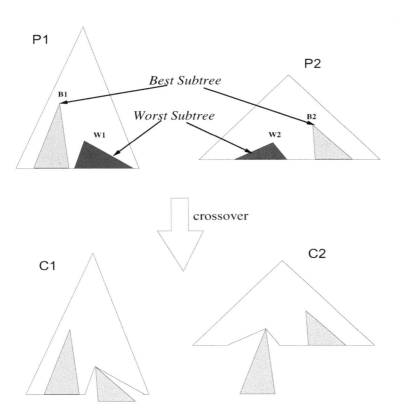

FIGURE 3.6: Crossover guidance [Iba *et al.*96a, Fig. 4].

This mechanism exploits already built structures (i.e., useful building-blocks) with adaptive recombination guided by MDL values.

We have confirmed the effectiveness of this guidance by experiments (see [Iba *et al.*96b] for details). Therefore, we believe STROGANOFF can guide GP recombination effectively in the sense that the recombination operation is guided using MDL values.

3.4   Numerical Problems with STROGANOFF

We applied STROGANOFF to several problems such as time series prediction, pattern recognition, and 0–1 optimization [Iba *et al.*93, Iba and Sato94b]. The results obtained were satisfactory. This section describes the experiments with time series predictions and compare the performance of STROGANOFF with other techniques.

3.4.1   Time Series Prediction with STROGANOFF

The Mackey–Glass differential equation

$$\frac{dx(t)}{dt} = \frac{ax(t - \tau)}{1 + x^{10}(t - \tau)} - bx(t), \qquad (3.36)$$

is used for time series prediction problems, where $a = 0.2$, $b = 0.1$, and $\tau = 17$ (Fig. 3.7(a)). This is a chaotic time series with a strange attractor of fractal dimension of approximately 3.5 [Tenorio *et al.*90].

In order to predict this series, the first 100 points (i.e., the values of $x(1), \cdots , x(100)$) were given to STROGANOFF as training data. The aim was to obtain a prediction of $x(t)$ in terms of M past data, i.e.,

$$x(t) = f(x(t - 1), x(t - 2), \cdots , x(t - M)). \qquad (3.37)$$

The parameters for STROGANOFF were as follows:

$N_{popsize}$	:	60
P_{cross}	:	0.6
P_{mut}	:	0.0333
T	:	$\{x(t - 1), x(t - 2), \cdots , x(t - 10)\}$

We used 10 past data for simplicity.

Fig. 3.8 shows the results of this experiment, namely the mean square error (S_N^2) and the MDL value as a function of the number of generations. Figs. 3.7(b) and (c) are the time series predicted by STROGANOFF (generations 233 and 1740, respectively). The MDL fitness values did not decrease monotonically, because the MDL values were plotted only when the minimum error-of-fit ratios improved. Note that the selection process of STROGANOFF

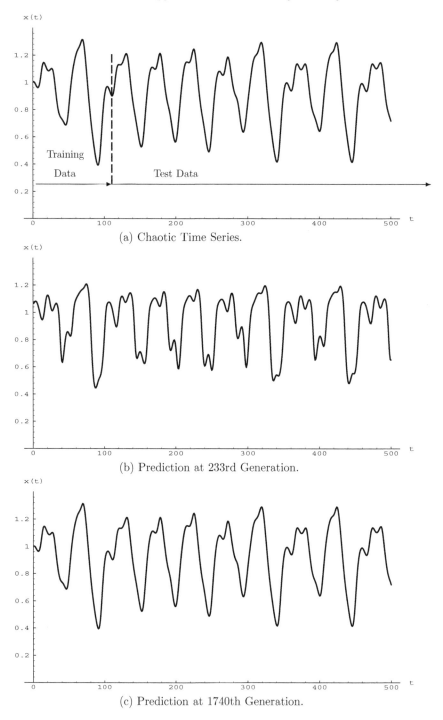

(a) Chaotic Time Series.

(b) Prediction at 233rd Generation.

(c) Prediction at 1740th Generation.

FIGURE 3.7: Predicting the Mackey–Glass equation [Iba *et al.*96a, Fig. 5].

is based on the MDL value, and not on the raw fitness (i.e., the error-of-fit ratio). The resulting structure of Fig. 3.7(c) was as follows:

```
(NODE95239 (7)
   (NODE95240
      (NODE95241
         (NODE95242
            (NODE95243
               (NODE95244
                  (8)
                  (NODE95245
                     (8)
                     (NODE95130 (2) (3))))
               (NODE95173
                  (10)
                  (NODE95174
                     (NODE95175 (4) (1))
                     (5)))))
         (5))
      (6))
   (NODE95178 (NODE95179 (8) (3)) (10))))
```

Where (i) represents $x(t - i)$. Some of the node coefficients were in Table 3.1. The mean square errors for this period are summarized in Table 3.2. The MDL-based fitness value of this tree is given as follows:

$$\begin{aligned}
\text{MDL fitness} &= 0.5k \log N + 0.5N \log S_N^2 \\
&= 0.5 \times (6 \times 13) \times \log 100 + 0.5 \times 100 \times \log(4.70 \times 10^{-6}) \\
&= -433.79.
\end{aligned}$$

Where the number of training data (i.e., N) is 100, and the MSE (i.e., S_N^2) is 4.70×10^{-6}. Since the number of intermediate nodes is 13, the k-value is roughly estimated as 6×13, because each internal node has six parameters.

Note that in Fig. 3.7(c) the prediction at the 1740th generation fit the training data almost perfectly. We then compared the predicted time series with the testing time series (i.e., $x(t)$ for $t > 100$). This also produced good results (compare Fig. 3.7(a) and Fig. 3.7(c)).

3.4.2   Comparison with a Traditional GP

Traditional GP has also been applied to the prediction task. In order to compare the performance of STROGANOFF, we applied a traditional GP system "sgpc1.1" (a Simple Genetic Programming in C written by Walter Alden

Table 3.1: Node coefficients [Iba *et al.*96a, Table 1].

Node	NODE95239	NODE95240	NODE95179
a_0	0.093	−0.090	0.286
a_1	0133	1.069	−0.892
a_2	0.939	−0.051	1.558
a_3	−0.029	1.000	1.428
a_4	0.002	−0.515	−0.536
a_5	−0.009	−0.421	−0.844

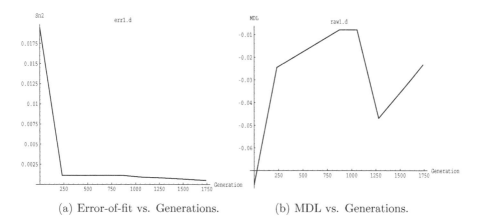

(a) Error-of-fit vs. Generations.   (b) MDL vs. Generations.

FIGURE 3.8: Time series prediction [Iba *et al.*96a, Fig. 6].

Table 3.2: Mean square errors (STROGANOFF) [Iba *et al.*96a, Table 2].

Generation	Training data	Testing data	MDL
233	0.01215	0.01261	−192.86
1740	$4.70{\times}10^{-6}$	$5.06{\times}10^{-6}$	−433.79

Table 3.3: GP parameters for predicting the Mackey–Glass equation.

Objective:	Predict next data $X(t)$ in Mackey–Glass mapping series.
Terminal set:	Time-embedded data series from $t = 1, 2, \cdots, 10.$, i.e., $\{X(t-1), X(t-2), \cdots, X(t-10)\}$, with a random constant.
Function set:	$\{+, -, \times, \%, \sin, \cos, \exp\}$.
Fitness cases:	Actual members of the Mackey–Glass mapping $(t = 1, 2, \cdots, 500)$.
Raw fitness:	Sum over the fitness cases of squared error between predicted and actual points.
Standardized fitness:	Same as raw fitness.
Parameters:	$M = 5000.$ $G = 101.$
Max. depth of new individuals:	6
Max. depth of mutant subtrees:	4
Max. depth of individuals after crossover:	17
Fitness-proportionate reproduction fraction:	0.1
Crossover at any point fraction:	0.2
Crossover at function points fraction:	0.7
Selection method:	Fitness-proportionate
Generation method:	Ramped half-and-half

Tackett) to the same chaotic time series (i.e., Mackey–Glass equation). For the sake of comparison, all the parameters chosen were the same as those used in the previous study [Oakley94, p.380, Table 17.3], except that the terminal set consisted of 10 past data for the short-term prediction (see Table 3.3).

Table 3.4 gives the results of the experiments, which show the mean square error of the best performance over 20 runs. For the sake of comparison, we also list the results given by STROGANOFF. The numbers of individuals to be processed are shown in the third column (i.e., #Pop×Gen).

The trees resulting from traditional GP are as follows:

Table 3.4: Mean square errors (GP vs. STROGANOFF) [Iba *et al.*96a, Table 4].

System	Gen.	#Pop.×Gen.	Training data	Testing data
STROGANOFF	233	13,980	0.01215	0.01261
	1,740	104,440	4.70×10^{-6}	5.06×10^{-6}
sgpc1.1(GP)	67	325,000	9.62×10^{-4}	2.08×10^{-3}
	87	435,000	6.50×10^{-6}	1.50×10^{-4}

```
<< Generation 67>>
(%
  (SIN
    (+
      (%
        X(t-8)
        X(t-3))
      X(t-9)))
  (%
    (%
      X(t-5)
      X(t-1))
    (%
      X(t-4)
      (EXP10
        X(t-7)))))
<< Generation 87>>
(-
  (+
    X(t-1)
    X(t-1))
  X(t-2))
```

The experimental results show that the traditional GP suffers from over-generalization, in the sense that the mean square error of the test data (i.e., 1.50×10^{-4}) is much worse than that of the training data (i.e., 6.50×10^{-6}). This may be caused by the fact that traditional GP has no appropriate criterion (such as MDL for STROGANOFF) for evaluating the trade-off between the errors and the model complexities (i.e., the description length of S-expressions).

Another disadvantage of traditional GP is due to the mechanisms used to generate constants. In traditional GP, constants are generated randomly by initialization and mutation. However, there is no tuning mechanism for the generated constants. This may degrade the search efficiency, especially in the case of time series prediction tasks, which require a fine-tuning of the fitting coefficients, so that the number of processed individuals for the same quality of solution is much greater than STROGANOFF (i.e., the third column in Table 3.4).

3.4.3   Statistical Comparison of STROGANOFF and a Traditional GP

In order to clarify these performance differences more statistically, we compare our approach with other prediction methods. More precisely, the predictor errors of the following techniques are compared using a variety of dynamic systems as case studies.

1. STROGANOFF

2. Traditional GP ("sgpc1.1" based on [Oakley94])

3. Radial basis functions [Franke82, Poggio and Girosi90]

3.4.3.1   Comparative Method

Given an m-dimensional chaotic time series,

$$\{x_1, \cdots, x_n, x_{n+1} \mid x_i \in \Re^m, \ x_{n+1} = f(x_n)\},$$

a predictor for x_{n+1} is described as follows:

$$x_{n+1} = \widetilde{f_N}(\{x_1, \cdots, x_n\}), \tag{3.38}$$

where N is the number of training data. Note that $\widetilde{f_N}$ is an m-dimensional vector function, i.e., $\widetilde{f_N} = (\widetilde{f_N^1}, \cdots, \widetilde{f_N^m})$. In order to quantify how well $\widetilde{f_N}$ performs as a predictor for f, the predictor error $\sigma^2(\widetilde{f_N})$ of $\widetilde{f_N}$ is defined by

$$\sigma^2(\widetilde{f_N}) = \lim_{M \to \infty} \frac{1}{M} \times \sum_{n=N}^{N+M-1} \|x_{n+1} - \widetilde{f_N}(\{x_n\})\|^2 / Var, \tag{3.39}$$

where

$$Var = \lim_{M \to \infty} M^{-1} \sum_{m=1}^{M} \|x_m - \lim_{M \to \infty} \sum_{m=1}^{M} x_m\|^2. \tag{3.40}$$

Var is a normalizing factor. $\| \cdot \|$ denotes the Euclidean norm on $\Re^m$. M is the number of test data and is set to be 10^3 in the following discussions.

In order to overcome certain higher-dimensional problems faced by STROGANOFF and traditional GP, we modify the previous predictor as follows. In the learning phase, we train the predictor $\widetilde{f_N^i}$ using the equation

$$x_t^i = \widetilde{f_N^i}(N(x_t)), \tag{3.41}$$

where $N(x_t)$ is the neighborhood of x_t. In the testing phase, we predict the future data x_{t+1} using the equation

$$x_{t+1}^i = \widetilde{f_N^i}(\{x_{j+1} \mid x_j \in N(x_t)\}). \tag{3.42}$$

x_{t+1} is derived using its neighborhood $N(x_{t+1})$ in the same way as the training phase. However, because $N(x_{t+1})$ is not known before x_{t+1} is derived, the predicted neighborhood $\{x_{j+1} \mid x_j \in N(x_t)\}$ is used as its substitute. The parameters used were the same as in the previous example, except that the terminal set included 10 past data in the neighborhood (i.e., $N(x_i)$). For instance, we used as the terminal set $\{x_1(t), \cdots, x_{10}(t), y_1(t), \cdots, y_{10}(t)\}$ for a two-dimensional problem, where $x_i(t)$ and $y_i(t)$ are the x- and y-coordinates of the i-th nearest past data to $x(t)$. We compare the performances of STROGANOFF and traditional GP (i.e., "sgpc1.1") with data estimated from "radial basis function predictors."

3.4.3.2 Comparison Tests

The following dynamic systems were chosen to compute predictor errors using the above techniques.

1. Ikeda map [Ikeda79]
   The Ikeda map is a two-dimensional dynamic system described below:

   $$f(x,y) = (1 + \mu(x \cos t - y \sin t), \quad \mu(x \sin t + y \cos t)), \tag{3.43}$$

   where

   $$t = 0.4 - \frac{6.0}{1 + x^2 + y^2}, \tag{3.44}$$

   $$\mu = 0.7. \tag{3.45}$$

   We consider the fourth iterate of the above Ikeda map (see Fig. 3.9):

   $$(x_{n+1}, y_{n+1}) = f^4(x_n, y_n). \tag{3.46}$$

   This map was chosen because it has a complicated functional form which is not of a type used in any of the above approximation techniques, but is well-behaved and slowly varying.

2. Lorenz equation [Lorenz63]
   Lorenz's equations consist of three simultaneous differential equations:

   $$\frac{dx(t)}{dt} = -3(x(t) - y(t)), \tag{3.47}$$

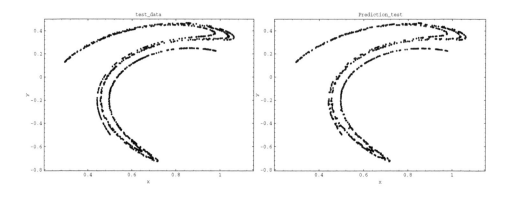

(a) Test Data.   (b) Prediction Result.

FIGURE 3.9: Ikeda map [Iba *et al.*96a, Fig. 7].

$$\frac{dy(t)}{dt} = -x(t)z(t) + 26.5x(t) - y(t), \qquad (3.48)$$

$$\frac{dz(t)}{dt} = x(t)y(t) - z(t), \qquad (3.49)$$

where
$$x(0) = z(0) = 0, \ y(0) = 1. \qquad (3.50)$$

We use sampling rates $\tau = 0.20$ (see Fig. 3.10).

3. Mackey–Glass equation [Mackey and Glass77]
   This delay differential equation was presented before, i.e.,

$$\frac{dx(t)}{dt} = \frac{ax(t-\tau)}{1+x^{10}(t-\tau)} - bx(t), \qquad (3.51)$$

The parameters used are

$$a = 0.2, \ b = 0.1, \ \tau = 17 \qquad (3.52)$$

3.4.3.3 Results

Figs. 3.9 and 3.10 show the test data and prediction results by STRO-
GANOFF, for the Ikeda map and the Lorenz equation, respectively. Table 3.5
shows the estimated values of $\log_{10} \sigma(\widetilde{f_N})$ for predictors $\widetilde{f_N}$ using three techniques. Also tabulated is the information dimension D. We conducted exper-
iments for STROGANOFF and traditional GP.

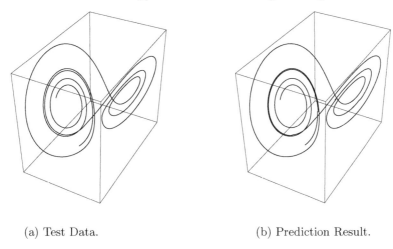

(a) Test Data.　　　　　　　　　　　(b) Prediction Result.

FIGURE 3.10: Lorenz attractor [Iba *et al.*96a, Fig. 8].

From the table, we see that the performance of traditional GP is very poor, especially in higher dimensional problems. We often observed that the best individuals acquired for the Ikeda map by traditional GP were simple expressions, shown below:

```
X(t) =
(-
   X1(t)
   (%
      X8(t)
      (EXP10
         (*
            (*
               X8(t)
               Y1(t))
            X7(t)))))

X(t) =
(+
   X1(t)
   (-
      Y8(t)
      Y8(t)))

X(t) = X1(t)

Y(t) = Y1(t)
```

Table 3.5: Estimated values of $\log_{10} \sigma(\widetilde{f_N})$ [Iba *et al.*96a, Table 5].

	D	N	Radial	GP	STROGANOFF
Ikeda	1.32	500	-2.10	-0.99	-1.23
Lorenz	2.0	500	-1.35	-0.55	-1.20
Mackey–Glass	2.1	500	-1.97	-1.43	-2.00

$Xi(t)$ and $Yi(t)$ are the x- and y-coordinates of the i-th nearest past data to $X(t)$. Note that the second expression is identical to $X1(t)$, because $X1(t) + (Y8(t) - Y8(t)) = X1(t) + 0 = X1(t)$. The first expression is also nearly equal to $X1(t)$, because the second term is close to zero (i.e., $\frac{X8(t)}{EXP10(X8(t) \times Y1(t) \times X7(t))} \approx 0$). $X1(t)$ and $Y1(t)$ are considered as very rough approximations of $X(t)$ in the training phase, because they are the closest points to $X(t)$ (see equation (3.41)). However, these are not effective predictors in the testing phase, in the sense that $X1(t)$ and $Y1(t)$ do not necessarily belong to the appropriate neighborhood of $X(t)$ (equation (3.42)). Therefore, the fact that the monomial expressions (i.e., $X1(t)$ and $Y1(t)$) often appeared in the resultant trees shows that traditional GP lacks generalization mechanisms, which in turn, results in the poor performance on the testing data.

The performance of STROGANOFF is by no means inferior to other techniques and gives acceptable results even in the case of the Ikeda map, which has complex dynamics (see Fig. 3.9).

Radial basis predictors seem superior to the other techniques. This technique is a global interpolation with good localization properties. It provides a smooth interpolation of scattered data in arbitrary dimensions and has proven useful in practice. However, with radial basis functions, the calculation of coefficients can be very costly for large N. On the other hand, STROGANOFF has several computational advantages described in Section 3.9.2.

The degree of the polynomial (the depth of the STROGANOFF tree) is adaptively tuned during the evolution of the trees. Therefore, we can conclude STROGANOFF offers an effective technique, which integrates a GP-based adaptive search of tree structures, and a local parameter tuning mechanism employing statistical search.

3.5   Classification Problems Solved by STROGANOFF

STROGANOFF has been applied to several classification tasks such as pattern recognition [Iba *et al.*93]. In this section, we show how STROGANOFF has generated an effective classifier to two pattern recognition tasks labelled "AYAME" (i.e., classifying whether a flower was an iris or not), and "SONAR"

Table 3.6: Data for pattern recognitions.

	Ayame	Sonar
# of Data	100	208
# of Features	4	60
Reference	[Fisher36]	[Gorman and Sejnowski88]

Table 3.7: GP parameters for classification.

Population size	60
Probability of graph crossover	0.6
Probability of graph mutation	0.0333
Terminal set	$\{(1), (2), (3), \cdots, (60)\}$

(i.e., classifying sonar echoes as coming from a submarine or not).
The training and testing data are summarized in Table 3.6.

3.5.1 AYAME Classification

Table 3.7 shows the parameters used. A terminal symbol (i) in the terminal set signifies the i-th feature. The final expression $\overline{y}$ was then passed through a threshold function.

$$\textbf{Threshold}(\overline{y}) = \begin{cases} 0, & \overline{y} < 0.5, \quad \text{negative example} \\ 1, & \overline{y} \geq 0.5, \quad \text{positive example} \end{cases} \qquad (3.53)$$

Fig. 3.11 shows the obtained results for Ayame's 2nd and 3rd classes, which plots the mean square error (S_N^2) and MDL value as a function of the number of generations. The acquired structures with node coefficients at generation 100 are shown in Tables 3.8 and 3.9. This tree classifies 100 data with 100% correctness.

Table 3.8: The best tree evolved for classification.

(NODE16347 (NODE16348 (NODE15772 (1) (NODE15773 (1) (NODE15774 (2) (2)))) (NODE16349 (2) (NODE16350 (4) (NODE16342 (NODE16343 (NODE16344 (4) (NODE16345 (NODE16346 (4) (4)) (2))) (1)) (NODE16121 (1) (NODE16122 (3) (NODE16123 (NODE16119 (2) (2)) (4))))))))) (NODE16091 (1) (NODE16092 (3) (NODE15911 (2) (4)))))

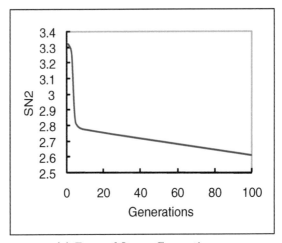

(a) Error-of-fit vs. Generations.

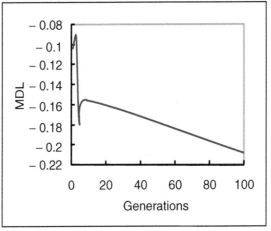

(b) MDL vs. Generations.

FIGURE 3.11: Pattern recognition (AYAME data).

Table 3.9: Node coefficients for AYAME data.

Node	NODE16347	NODE16348	NODE15911
a_0	-0.021	-0.021	-0.524
a_1	1.980	0.080	-0.444
a_2	-0.884	0.977	1.573
a_3	-8.582	-0.148	0.770
a_4	3.673	-0.027	-0.217
a_5	4.815	0.099	-0.763

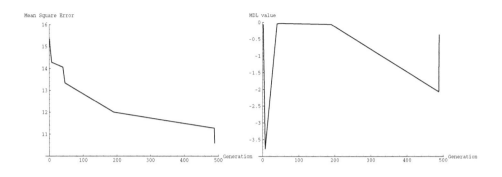

(a) Error-of-fit vs. Generations.   (b) MDL vs. Generations.

FIGURE 3.12: Pattern recognition (SONAR data).

3.5.2 SONAR Classification

Next, we present the results of an experiment called "SONAR" (i.e., classifying sonar echoes as coming from a submarine or not). In the "SONAR" experiment, the number of data points was 208 and the number of features was 60 [Gorman and Sejnowski88]. We used the same parameters shown in Table 3.7.

We randomly divided 208 data points into two groups, 168 training data and 40 testing data.

Fig. 3.12 shows the results, i.e., mean square error and MDL vs. the number of generations. The elite evolved structure at the 489th generation classified the training data with 92.85% accuracy, and the testing data with 78.94% accuracy. This result is comparable to that of [Gorman and Sejnowski88].

The resulting elite structure evolved at the 489th generation is shown in Tables 3.10 and 3.11.

To compare the performance of STROGANOFF with those of neural networks, we conducted an experiment using the same conditions as described in [Gorman and Sejnowski88]. The combined set of 208 cases was divided randomly into 13 disjoint sets, so that each set consisted of 16. For each experiment, 12 of these sets were used as training data, while the 13th was reserved for testing. The experiment was repeated 13 times in order for every case to appear once as a test set. The reported performance was obtained by averaging the 13 different test sets, with each set run 10 times.

The results of this experiment are shown in Table 3.12. The table shows averages and standard deviations of accuracy for test and training data. NN (i) indicates a neural network with a single hidden layer of i hidden units. The neural data was reported by Gorman and Sejnowski [Gorman and Sejnowski88].

Table 3.10: The best tree evolved for classification.

(NODE35040 (NODE35041 (54) (NODE35042 (NODE35043 (34) (NODE35044 (22) (NODE35045 (NODE35046 (NODE35047 (46) (NODE35048 (40) (NODE34446 (39) (32)))) (NODE34650 (NODE-34623 (NODE34624 (NODE34625 (36) (NODE34626 (NODE34627 (NODE34628 (35) (NODE34629 (NODE34630 (42) (20)) (NODE-34631 (48) (NODE34632 (NODE34633 (52) (23)) (48))))) (NODE34634 (NODE34635 (NODE34636 (28) (24)) (41)) (NODE34637 (9) (17)))) (NODE34638 (NODE34639 (NODE34640 (17) (43)) (NODE34641 (26) (38))) (NODE34642 (1) (30))))) (31)) (40)) (26))) (NODE34667 (NODE34668 (44) (NODE34453 (52) (52))) (NODE34454 (NODE-34455 (NODE34456 (NODE34457 (NODE34458 (60) (NODE34459 (2) (NODE34460 (NODE34461 (59) (NODE34462 (4) (NODE34463 (36) (42)))) (4)))) (NODE34464 (NODE34465 (54) (NODE34466 (28) (14))) (NODE34467 (48) (NODE34468 (NODE34469 (NODE34470 (51) (22)) (13)) (NODE34471 (NODE34472 (NODE34473 (34) (NODE34474 (46) (47))) (57)) (20)))))) (NODE34475 (NODE34476 (52) (NODE-34477 (53) (20))) (NODE34478 (42) (3)))) (NODE34479 (NODE34480 (NODE34481 (29) (NODE34482 (NODE34483 (NODE34484 (58) (58)) (19)) (NODE34485 (58) (35)))) (NODE34486 (57) (NODE34487 (16) (NODE34488 (41) (NODE34489 (NODE34490 (NODE34491 (31) (43)) (37)) (NODE34492 (18) (16))))))) (NODE34493 (47) (14)))) (NODE-34494 (23) (NODE34495 (57) (23))))))))) (6))) (NODE34151 (NODE-33558 (NODE33548 (NODE33549 (50) (49)) (25)) (56))

The following points from the table should be noticed.

1. STROGANOFF is not necessarily as good as neural networks at learning training data.

2. Judging from the neural testing data results, neural networks may suffer from an over-fitting of the training data.

3. STROGANOFF established an adequate generalization resulting in good results for the testing data.

Table 3.11: Node coefficients for SONAR data.

Node	NODE34152	NODE33549	NODE33548
a_0	1.484821	0.89203346	0.27671432
a_1	-3.2907495	2.631569	0.79503393
a_2	-12.878513	-14.206657	-0.66170883
a_3	1.5597038	-256.3501	0.95752907
a_4	3.294608	275.04077	-0.395895
a_5	55.61087	111.96045	0.11753845

Table 3.12: STROGANOFF vs. neural networks.

	Training data		Testing data	
	Avg.	Std.	Avg.	Std.
STROGANOFF	88.9	1.7	85.0	5.7
NN (0)	89.4	2.1	77.1	8.3
NN (2)	96.5	0.7	81.9	6.2
NN (3)	98.8	0.4	82.0	7.3
NN (6)	99.7	0.2	83.5	5.6
NN (12)	99.8	0.1	84.7	5.7
NN (24)	99.8	0.1	84.5	5.7

Thus we conclude that the MDL-based fitness functions can be used for effective control of tree growth. These methods appear to achieve good results for the testing data by not fitting the training data in a strict manner.

3.6 Temporal Problems Solved by STROGANOFF

In this section, we extend STROGANOFF to cope with temporal events and establish a new system $\Re$-STROGANOFF (Recurrent STROGANOFF). $\Re$-STROGANOFF integrates a GP-based adaptive search of tree structures, and a parameter tuning mechanism employing an error-propagation method. We demonstrate the effectiveness of our new system with several experiments.

3.6.1 Introducing Recurrency in STROGANOFF

This section explains how we extended the STROGANOFF algorithm to be able to cope with temporal data processing, by using special "memory terminals", which point to any non-terminal node within the tree. The input of this memory terminal is the past output of the non-terminal node pointed to by the memory terminal. Fig. 3.13 shows an exemplar tree with "memory terminals." This tree can be written as an S-expression in LISP,

```
(NODE1
    (NODE2
        (NODE3 (x₁) (x₂))
        (M_NODE2)
    (M_NODE1)))
```

where x_1 and x_2 are the input variables. M_NODE1 and M_NODE2 are memory terminals, which point to NODE1 and NODE2, respectively. Intermediate nodes represent simple polynomial relationships between two descendant (lower) nodes. Assuming that we use quadratic expressions, the following equations are given at the current time t:

$$\text{NODE3: } z_1(t) = a_0 + a_1 x_1(t) + a_2 x_2(t) + a_3 x_1(t) x_2(t)$$
$$+ a_4 x_1(t)^2 + a_5 x_2(t)^2, \tag{3.54}$$

$$\text{NODE2: } z_2(t) = b_0 + b_1 z_1(t) + b_2 z_2(t-1) + b_3 z_1(t) z_2(t-1)$$
$$+ b_4 z_1(t)^2 + b_5 z_2(t-1)^2, \tag{3.55}$$

$$\text{NODE1: } \overline{y}(t) = c_0 + c_1 z_2(t) + c_2 \overline{y}(t-1) + c_3 z_2(t) \overline{y}(t-1)$$
$$+ c_4 z_2(t)^2 + c_5 \overline{y}(t-1)^2, \tag{3.56}$$

where $x_1(t)$ and $x_2(t)$ are inputs of x_1 and x_2 at time t. M_NODE2 and M_NODE1 are replaced by $z_2(t-1)$ and $\overline{y}(t-1)$.

Initially, all coefficients $(a_i, b_i, c_i; i = 0, \cdots, 5)$ are randomly chosen. At each time step, they are updated according to the rules derived from error back-propagation (BP, see Section 3.6.2 for details). For instance, we have the update rule for c_0 at time t as follows:

$$\Delta c_0(t) = -\alpha \frac{\partial \overline{y}(t)}{\partial c_0} \left(\overline{y}(t) - y(t) \right), \tag{3.57}$$

where $y(t)$ is the desired output value at time t, and α is some fixed positive learning rate. With the initial conditions ($\frac{\partial \overline{y}(0)}{\partial c} = \frac{\partial z_2(0)}{\partial c} = 0$), $\frac{\partial \overline{y}(t)}{\partial c_0}$ is determined from the following recursive equations:

$$\frac{\partial \overline{y}(t)}{\partial c_0} = 1 + (c_1 + c_3 \overline{y}(t-1) + 2c_4 z_2(t)) \frac{\partial z_2(t)}{\partial c_0}$$
$$+ (c_2 + c_3 z_2(t) + 2c_5 \overline{y}(t-1)) \frac{\partial \overline{y}(t-1)}{\partial c_0} \tag{3.58}$$

$$\frac{\partial z_2(t)}{\partial c_0} = (b_2 + b_3 z_1(t) + 2b_5 z_2(t-1)) \frac{\partial z_2(t-1)}{\partial c_0} \tag{3.59}$$

The above weight adjustment is an extension of error back-propagation commonly used in neural networks (see [Williams and Zipser89] for instance). We call this process "GEP" (generalized error-propagation). The main differences between GEP and usual BP are summarized as follows:

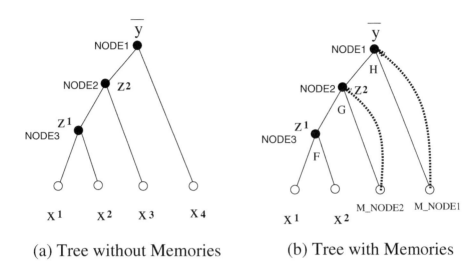

(a) Tree without Memories

(b) Tree with Memories

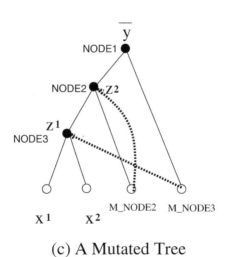

(c) A Mutated Tree

FIGURE 3.13: Exemplar tree for ℜ-STROGANOFF.

1. Whereas BP uses a logistic function of linear sum of inputs, GEP handles an arbitrary polynomial. Moreover we can define a different function type for each non-terminal node.

2. In usual BP, recurrency is used only as a "second-order" connection (i.e., the recurrent connections allow the network's hidden units to see their own previous outputs). In our approach, a "memory terminal" can point to any intermediate node. Therefore, we have to derive recursive equations following the paths of the tree in order to propagate errors.

3. We use polynomials of two variables, because the trees in the GP genotype are binary. This restriction limits the number of branching factors in recursion formula, which in turn reduces the computational cost.

With these considerations, our ℜ-STROGANOFF algorithm is as follows:

Step1 Initialize a population of tree expressions and calculate all coefficients using GEP.

Step2 Evaluate each expression in the population so as to derive the MDL-based fitness.

Step3 Create new expressions (children) by mating current expressions. Apply mutation and crossover to the parent tree expressions.

Step4 Replace the members of the population with the child trees.

Step5 If any node recurrency in a child tree is changed as a result of recombination, the GEP process is executed, so as to compute the coefficients of the nodes of the tree.

Step6 If the termination criterion is satisfied, then halt; else go to **Step2**.

We use a Minimum Description Length (MDL)-based fitness definition for evaluating the tree structures, i.e., equations (3.32)–(3.35). In **Step5**, GEP is executed only when a "memory terminal" happens to point to a different (i.e., mutated) intermediate node as a result of recombination. For instance, consider the tree shown in Fig. 3.13(a). If the subtree below NODE3 is changed in some way, the recurrencies M_NODE1 and M_NODE2 are changed so as to recalculate the coefficients $(b_0, b_1, \cdots, c_5)$. On the other hand, if M_NODE1 is mutated to point to NODE3 (Fig. 3.13(b)), only the coefficients of NODE1 should be changed. The coefficients of NODE2 and NODE3 are unchanged, because the subtree below NODE2 is unchanged and so is the recurrency M_NODE2. This reduces the number of the repetition of the GEP process, as the generations proceed. Although this restriction in the number of GEP repetition has no theoretical background, we will see that ℜ-STROGANOFF shows satisfactory performance in the following section.

3.6.2 GEP (Generalized Error-Propagation) Algorithm

We define the error at time t to be

$$E(t) = \frac{1}{2} \left(\overline{y}(t) - y(t) \right)^2, \tag{3.60}$$

where $y(t)$ is a desired output value at time t. Any coefficient w should be updated with a gradient-descent weight update rule:

$$\Delta w = -\alpha \frac{\partial E(t)}{\partial w}, \tag{3.61}$$

where α is some fixed positive learning rate. Thus we want to compute

$$\frac{\partial E(t)}{\partial w} = \frac{\partial \overline{y}(t)}{\partial w} \left(\overline{y}(t) - y(t) \right). \tag{3.62}$$

Since the values of $\overline{y}(t)$ and $y(t)$ are known at time t, all that remains is to compute $\frac{\partial \overline{y}(t)}{\partial w}$. This can be done by calculating the recursive equations following the paths in a given tree.

Consider the tree shown in Fig. 3.13(a). For the sake of simplicity, we use the following equations:

$$\text{NODE3: } z_1(t) = H(h_1, h_2)$$
$$= a_0 + a_1 h_1 + a_2 h_2 + a_3 h_1 h_2 + a_4 h_1^2 + a_5 h_2^2, \tag{3.63}$$

$$\text{NODE2: } z_2(t) = G(g_1, g_2)$$
$$= b_0 + b_1 g_1 + b_2 g_2 + b_3 g_1 g_2 + b_4 g_1^2 + b_5 g_2^2, \tag{3.64}$$

$$\text{NODE1: } \overline{y}(t) = F(f_1, f_2)$$
$$= c_0 + c_1 f_1 + c_2 f_2 + c_3 f_1 f_2 + c_4 f_1^2 + c_5 f_2^2, \tag{3.65}$$

where $h_1 = x_1(t)$, $h_2 = x_1(t)$, $g_1 = z_1(t)$, $g_2 = z_2(t-1)$, $f_1 = z_2(t)$, $f_2 = \overline{y}(t-1)$. We now obtain the following recursive expressions:

$$\frac{\partial \overline{y}(t)}{\partial c} = F_c + F_{f_1} \frac{\partial f_1}{\partial c} + F_{f_2} \frac{\partial f_2}{\partial c}$$
$$= F_c + F_{f_1} \frac{\partial z_2(t)}{\partial c} + F_{f_2} \frac{\partial \overline{y}(t-1)}{\partial c},$$

where $F_c = \frac{\partial F(f_1, f_2)}{\partial c}$, $F_{f_1} = \frac{\partial F(f_1, f_2)}{\partial f_1}$, and $F_{f_2} = \frac{\partial F(f_1, f_2)}{\partial f_2}$. c is a coefficient of F, i.e., $c = c_0, c_1, \cdots, c_5$.

Table 3.13: $\Re$-STROGANOFF parameters.

Population size	120
Crossover probability	60%
Mutation probability	3.3%
Terminal node	$x(t)$ (input at time t)
Non-terminal node (subexpression)	$F(x,y) = a_0(1-x)(1-y)$ $+a_1 x(1-y) + a_2(1-x)y + a_3 xy$ where $0 \leq a_0, a_1, a_2, a_3 \leq 1$

In order to obtain the value of $\frac{\partial z_2(t)}{\partial c}$, we expand the recurrent relation further as follows:

$$\frac{\partial z_2(t)}{\partial c} = G_{g_1}\frac{\partial g_1}{\partial c} + G_{g_2}\frac{\partial g_2}{\partial c} = G_{g_1}\frac{\partial z_1(t)}{\partial c} + G_{g_2}\frac{\partial z_2(t-1)}{\partial c}, \tag{3.66}$$

$$\frac{\partial z_1(t)}{\partial c} = H_{h_1}\frac{\partial h_1}{\partial c} + H_{h_2}\frac{\partial h_2}{\partial c} = H_{h_1}\frac{\partial x_1(t)}{\partial c} + H_{h_2}\frac{\partial x_2(t)}{\partial c}$$
$$= H_{h_1} \times 0 + H_{h_2} \times 0 = 0. \tag{3.67}$$

Thus we have

$$\frac{\partial z_2(t)}{\partial c} = G_{g_2}\frac{\partial z_2(t-1)}{\partial c}. \tag{3.68}$$

Therefore we can derive $\frac{\partial \bar{y}(t)}{\partial c}$ at time t, with the initial conditions ($\frac{\partial \bar{y}(0)}{\partial c} = \frac{\partial z_2(0)}{\partial c} = 0$).

Other derivatives for the weight updates, such as $\frac{\partial \bar{y}(t)}{\partial b_i}$ and $\frac{\partial \bar{y}(t)}{\partial a_i}$ ($i = 0, 1, \cdots, 5$), are derived in a similar way.

3.6.3 Experimental Results

This section shows the effectiveness of $\Re$-STROGANOFF by using the results of several experiments. For the sake of comparison, most of these experiments were taken from previous papers. We used the parameters shown in Table 3.13 throughout these experiments.

The reason we chose a bilinear function (i.e., $F(x,y)$) is as follows:

1. If $0 \leq x, y \leq 1$, then $0 \leq F(x,y) \leq 1$. Thus, a divergence problem of outputs does not occur.

2. The resultant expression is often interpretable as a simple Boolean function. For instance, if $a_0 = a_1 = a_2 = 0, a_3 = 1$, then $F(x,y) = x \wedge y$.

Table 3.14: Node coefficients of the acquired tree.

Node	Coefficients			
	a_0	a_1	a_2	a_3
NODE5071168	0.0	1.0	0.395	0.484
NODE5070912	1.0	0.314	0.130	0.0
NODE5071104	0.0	0.0	0.228	0.567
NODE5071040	0.389	0.914	0.812	0.00015
NODE5070976	0.462	0.704	0.377	0.832

If $a_0 = 0, a_1 = a_2 = a_3 = 1$, then $F(x, y) = x \vee y$. STROGANOFF also used this function for Boolean learning as described in the previous section.

3.6.3.1 Learning to Oscillate

The first experiment was to oscillate, because oscillation is an interesting class of behaviors to study with any algorithm designed to train arbitrary network dynamics [Williams and Zipser89]. The target oscillation is a simple sequence alternating 0's and 1's. $\Re$-STROGANOFF showed satisfactory results for various oscillation tasks. For instance, when the sequence

```
0011001100110011......
```

was given, the following tree was acquired at the generation of 16.

```
(NODE5071168
  (NODE5070912
    M_NODE5071168
    M_NODE5071168)
  (NODE5071104
    (NODE5071040
      M_NODE5071104
      (NODE5070976
        X(t)
        M_NODE5070976))
    M_NODE5071104))
```

where $x(i) = 0, x(i + 1) = 0, x(i + 2) = 1, x(i + 3) = 1$ for $i = 0, 1, 2, \cdots$. The node coefficients are shown in Table 3.14. This tree gives the exact oscillating sequence (i.e., $00110011\cdots$). Fig. 3.14 shows the best of each generation of evolved oscillations at the generations of 0,4,10, and 16.

3.6.3.2 Inducing Languages from Examples

The second experiment was to induce finite-state languages over $\{0, 1\}^*$ from examples. These languages were selected for study by [Tomita82] (see

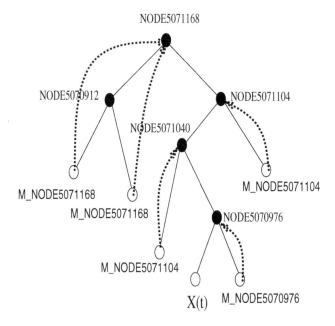

FIGURE 3.14: An evolved tree for the oscillation task.

Table 3.15). Tomita also selected for each language a set of positive and negative examples to be used as a training set. Following [Watrous and Kuhn92], we used the Tomita training set of grammatical strings of length 10 or less. The test set consisted of all strings of length 10 or less.

A training cycle consists of presenting a whole set of examples. For instance, in the case of L1, the training cycle includes 17 ($= 9 + 8$) presentations of the examples. At each time step, one symbol from the string is presented to ℜ-STROGANOFF. After the final symbol of each string is presented, the desired output is given, i.e., "1" for positive examples or "0" for negative examples. Notice that training (i.e., the GEP process) occurs after each string presentation. The GEP process is executed for each tree until it classifies all examples correctly or until a maximum number of cycles (i.e., 100) is reached.

For the sake of classification, the final expression $\overline{y}$ was then passed through a threshold function.

$$\mathbf{Threshold}(\overline{y}) = \begin{cases} 0, & \overline{y} < 0.5, \quad \text{negative example} \\ 1, & \overline{y} \geq 0.5, \quad \text{positive example} \end{cases} \tag{3.69}$$

The following tree was acquired at the generation of 6 for L1:

```
(NODE223080
    (NODE219344
        M_NODE223080
```

Table 3.15: Tomita languages.

Lang.	Definition			
	# of strings (Len. ≤ 10)		# of training data	
	Positive	Negative	Positive	Negative
L1	1*			
	11	2036	9	8
L2	(10)*			
	6	2041	5	10
L3	no odd-length 0-string anywhere after an odd-length 1-string			
	716	1331	11	11
L4	any string not containing 000 as a substring			
	1103	944	10	7

Table 3.16: Coefficients of the acquired tree (2).

Node	Coefficients			
	a_0	a_1	a_2	a_3
NODE223080	0.0	0.0	0.999	0.00016
NODE219344	1.0	0.0	0.00397	0.0
NODE216576	0.0	1.0	0.0	1.0
NODE216960	0.306	0.618	0.213	0.500
NODE218760	0.528	0.557	0.207	0.447

```
   M_NODE223080)
 (NODE216576
   X(t)
   (NODE216960
     X(t)
     (NODE218760
       M_NODE218760
       M_NODE218760))))
```

The acquired coefficients are shown in Table 3.16.

This tree classified all the test data with 100% accuracy, i.e., the evolved tree classified all unseen strings.

The best tree evolved for L4 at generation 1299 is shown in Table 3.17. The evolved tree classified the test set of strings up to length 5 with 100% accuracy. The accuracy for the test set of strings up to length 10 was 91.2%. This degradation is due to the possibility of error accumulation as the tree classifies long, unseen strings.

Table 3.17: The best tree evolved for L4.

(NODE309184 (NODE309320 (NODE309456 (NODE309592 (NODE309728   (NODE309864   X(t)   M_NODE309320) M_NODE312408) X(t)) (NODE310208 (NODE310344 (NODE-310480 M_NODE309320 X(t)) (NODE310720 M_NODE312408 (NODE310928 X(t) M_NODE309320))) (NODE311168 (NODE-311304 M_NODE312408 (NODE311512 X(t) M_NODE309320)) (NODE311752 (NODE311896 X(t) M_NODE309320) X(t))))) (NODE312168 M_NODE312408 X(t))) (NODE312408 (NODE-312544 (NODE312680 (NODE312816   X(t)   (NODE312984 (NODE313120 M_NODE312408 X(t)) (NODE313360 (NODE-313496 X(t)   M_NODE309320) (NODE313736 M_NODE-312408 M_NODE309320)))) (NODE314016 M_NODE312408 M_NODE309320)) X(t)) (NODE314328 (NODE314464 (NODE-314600 M_NODE312408 M_NODE309320) M_NODE312408) M_NODE309320)))

We experimented with other languages as shown in the table, and got satisfactory results.

3.6.3.3   Extracting FSA (Finite-State Automata)

Following [Giles *et al.*92], we experimented in extracting what the ℜ-STRO-GANOFF tree learned. This FSA extraction process includes the following steps: (1) Clustering FSA states, (2) Constructing a transition diagram by connecting these states together with the alphabet (i.e., 0 or 1) labelled arcs. The clusters of states are formed by dividing each memory terminal's output range $[0, 1]$ into q partitions of equal width. A "memory node" is an intermediate node pointed to by some memory terminal. Thus for M "memory terminals", there exist q^M possible states.

Consider the evolved tree for L4 in Table 3.17. We set $q = 2$. In this tree, there are two "memory terminals" (NODE309320 and NODE312408), i.e., $M = 2$. Therefore we have $2^2 = 4$ FSA states. As a result of the extraction described above, we obtained the FSA shown in Fig. 3.15. This automaton is an ideal FSA for L4. What is interesting is that this FSA correctly classifies all unseen strings whereas the evolved tree does not.

3.6.4   Comparison with Other Methods

The experiments discussed above show that our ℜ-STROGANOFF approach is an effective strategy for temporal data processing. We feel that ℜ-STROGANOFF searches in an efficient manner, because it integrates an error-propagation method and a GP-based evolutionary search strategy.

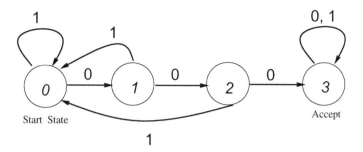

FIGURE 3.15: An evolved automaton for L4.

Table 3.18: Performance comparison.

Lang.	Best Accuracy	Method [Reference]
L1	100.0%	recurrent neural networks [Watrous and Kuhn92, p.410]
L1	100.0%	ℜ-STROGANOFF
L4	60.92%	recurrent neural networks [Watrous and Kuhn92, p.410]
L4	100.0%	recurrent neural networks [Giles *et al.*92, p.401]
L4	91.2%	ℜ-STROGANOFF

The error-propagation method is commonly used in neural networks. Thus it was thought to be instructive to compare the performance of ℜ-STRO-GANOFF to those of neural networks, as shown below. The comparison of performances for Tomita languages L1 and L4 is shown in Table 3.18.

Table 3.18 shows the best performances reported in the literature[2]. As can be seen, ℜ-STROGANOFF works almost as well as recurrent neural networks. Further analysis of the performance of ℜ-STROGANOFF remains to be seen.

3.7 Financial Applications by STROGANOFF

There have been several applications of GA or GP to the financial tasks, such as portfolio optimization, bankruptcy prediction, financial forecasting,

---

[2][Giles *et al.*92] used a more complicated training scheme for the presentation of samples. The testing data consisted of all strings up to length 15.

fraud detection and portfolio selection [Aranha and Iba08]. GP has also been successfully applied in the generation of algorithmic trading systems [Potvin *et al.*04, Aranha *et al.*07].

We show how the decision rule derived by STROGANOFF can successfully predict the stock pricing, and gain high profits in a market simulation. Comparative experiments are conducted with standard GP and neural networks to show the effectiveness of our approach.

3.7.1　Predicting Stock Market Data

In this section, we present the application of STROGANOFF to predict a real-world time series, i.e., the prediction of the price data in the Japanese stock market. Our goal is to make an effective decision rule as to when and how many stocks to deal, i.e., sell or buy. Our method is used to predict the price data in Japanese stock market. The financial data we use is the stock price average of Tokyo Stock Exchange, which is called Nikkei225.

Evolutionary algorithms have been applied to the time series prediction, such as sunspot data [Angeline96] (see also Section 2.2.3) or the time sequence generated from the Mackey–Glass equation (Section 3.4.1). Among them, the financial data prediction provides a challenging topic. This is because the stock market data are quite different from other time series data for the following reasons:

1. The ultimate goal is not to minimize the prediction error, but to maximize the profit gain.

2. Stock market data are highly time-variant.

3. The stock market data are given in an event-driven way. They are highly influenced by the indeterminate dealing.

3.7.1.1　Target Financial Data

The Nikkei225 average is computed by the Nihon Keizai Shimbun-Sha, a well-known financial newspaper publishing firm. The derivation is based upon the Dow formula. As of April 16th 2009, the Nikkei average stood at 8,981.90 Japanese yen (JPY). However, this average is a theoretical number and should be rigidly distinguished from the real average price in the market place. The computation formula for the Nikkei average (Nikkei22) is as follows:

$$\text{Nikkei Average} = \frac{\sum_{x \in 225 \text{ stocks}} \text{Price}_x}{D} \tag{3.70}$$

The sum of the stock price, Price_x is over 225 representative stocks in Tokyo Stock Exchange market. Originally, the divisor D was 225, i.e., the number of component stocks. However, the divisor is adjusted whenever price changes resulting from factors other than those of market activity take place. The

Nikkei averages are usually given every minute from 9:00 am to 12:00 pm and from 1:00 pm to 3:00 pm. The data we use in the following experiments span over a period from April 1st 1993 to September 30th 1993. Fig. 2.21 shows the example tendency of the Nikkei225 average (tick data) during the above period. All data are normalized between 0.0 and 1.0 as the input value. The total number of data is 33,177. We use the first 3,000 time steps for the training data and the rest for the testing data.

3.7.1.2 STROGANOFF Parameters and Experimental Conditions

We have applied STROGANOFF to predicting the Nikkei225 stock price average. The used parameters are shown in Table 3.19. For the sake of comparison, STROGANOFF was run using a variety of terminal sets described below.

- Condition A: The terminal set is $\{y1, \cdots, y10, \Re\}$, in which yi is the Nikkei225 price average observed i minutes before the predicted time. That is, if $x(t)$ is the Nikkei225 price average at time t, then $yi = x(t-i)$. $\Re$ is a constant generated randomly.

- Condition B: The terminal set is $\{ave1, \cdots, ave10, \Re\}$. The $avei$ terminal is the moving average of the Nikkei225 value every 10 minutes, i.e.,

$$ avei = \frac{\sum_{k=1}^{10} x(t - 10 * (i - 1) - k)}{10}. $$

- Condition C: The terminal set is $\{m1, \cdots, m10, \Re\}$. The mi terminal is the variance of the Nikkei225 value every 10 minutes, i.e.,

$$ mi = \frac{\sum_{k=1}^{10} (x(t - 10 * (i - 1) - k) - avei)^2}{10}. $$

- Condition D: The terminal set is $\{m1, \cdots, m10, ave1, \cdots, ave10, \Re\}$.

- Condition E: The terminal set is $\{v1, \cdots, v10, r1, \cdots, r10, \Re\}$, where the terminals vi and ri are defined as follows:

$$ vi = |x(t - i) - x(t - i - 1)| $$

$$ ri = \frac{x(t - i) - x(t - i - 1)}{x(t - i - 1)}. $$

The predicted value, i.e., the target output of a STROGANOFF tree, is the current Nikkei225 price average for the conditions from A to D. On the other hand, for the condition E, the target is the difference between the current Nikkei225 price average and the price observed one minute before. The mean square error is derived from the predicted value and the target data. Then, the fitness value is calculated as follows:

Table 3.19: STROGANOFF parameters.

max_generation	100	max_depth_after_crossover	17
population_size	100	max_depth_for_new_trees	6
steady_state	0	max_mutant_depth	4
grow_method	GROW	crossover_any_pt_fraction	0.2
tournament_K	6	crossover_func_pt_fraction	0.7
selection_method	TOURNAMENT	fitness_prop_repro_fraction	0.1
weigh value w	$w \in \{0.2, 0.1, 0.01, 0.001, 0.0001, 0.0, -0.01\}$		

$$\text{MDL fitness} = 0.5kW \log N + 0.5N \log S_N^2, \qquad (3.71)$$

where where N is the number of input–output data pairs, S_N^2 is the mean square error. In this equation, we modified the previous definition of MDL (equation (3.32)) so as to use the weight value W.

3.7.1.3   GP Parameters and Experimental Conditions

For the sake of comparison, standard GP was also applied to the same data. We chose sgpc1.1, a simple GP system in C language, for predicting the Nikkei225 stock price average. The used parameters are shown in Table 3.20. GP was run using the same terminal sets as those used by STROGANOFF (see Section 3.7.1.2).

The GP fitness value is defined to be the mean square error of the predicted value and the target data. The smaller the fitness value, the better.

3.7.1.4   Validation Method

In order to confirm the validness of the predictor acquired by STROGANOFF and GP, we examine the best evolved tree with the stock market simulation during the testing period. Remember that the output prediction of a tree is the current Nikkei225 price average for conditions from A to D. Thus, we use the following rule to choose the dealing, i.e., to decide whether to buy or sell a stock. Let $Pr(t)$ be the observed Nikkei225 average at the time step of t.

Step1 Initially, the total budget BG is set to be 1,000,000 JPY (Japanese yen). Let the time step t be 3,000, i.e., the beginning of the testing period. The stock flag ST is set to be 0.

Step2 Derive the output, i.e., the predicted Nikkei225 average, of the GP tree. Let $\widetilde{Pr}(t)$ be the predicted value.

Step3 If $Pr(t-1) < \widetilde{Pr}(t)$ and $ST = 0$, then buy the stock. That is, set ST to be 1.

Step4 Else, if $Pr(t-1) > \widetilde{Pr}(t)$ and $ST = 1$, then sell the stock. That is, set ST to be 0.

Step5 If $ST = 1$, let $BG := BG + Pr(t) - Pr(t-1)$.

Step6 If $BG < 0$, then return 0 and stop.

Step7 If $t < 33,177$, i.e., the end of the testing period, then $t := t+1$ and go to **Step2**. Else return the total profit, i.e., $BG-1,000,000$ JPY.

The stock flag ST indicates the state of holding stock, i.e., if $ST = 0$, then no stock is shared at present, whereas if $ST = 1$, then a stock is shared. In **Step5**, the total property is derived according to the newly observed stock price. The satisfaction of the **Step6** condition means that the system has gone into bankruptcy.

For the condition E, the tree outputs the difference between the current Nikkei225 price average and the price observed one minute before. Let the predicted output be $\widetilde{Pr'}(t)$. Then the dealing condition depends on the output value itself. More precisely, the above steps are revised as follows:

Step3 If $0 < \widetilde{Pr'}(t)$ and $ST = 0$, then buy the stock. That is, set ST to be 1.

Step4 Else, if $0 > \widetilde{Pr'}(t)$ and $ST = 1$, then sell the stock. That is, set ST to be 0.

We use the above dealing rules for the validation of the acquired STRO-GANOFF or GP tree. For the sake of simplicity, we put the following assumptions on the market simulation:

1. At most one stock is shared at any time.

2. The dealing stock is imaginary, in the sense that its price behaves exactly the same as the Nikkei225 average price.

The optimal profit according to the above dealing rule is 80,106.63 yen. This profit is ideally gained when the prediction is perfectly accurate during the testing period.

3.7.1.5 Experimental Results

STROGANOFF and GP runs were repeated under each condition 10 times. The training and the validation performance is shown in Tables 3.21 and 3.22. The MSE values are the average of mean square errors given by the best evolved tree for the training data. The hit percentage means how accurately the GP tree made an estimate of the qualitative behavior of the price. That is, the hit percentage is calculated as follows:

Table 3.20: GP parameters for sgpc1.1.

max_generation	100	max_depth_after_crossover	17
population_size	1000	max_depth_for_new_trees	6
steady_state	0	max_mutant_depth	4
grow_method	GROW	crossover_any_pt_fraction	0.2
tournament_K	6	crossover_func_pt_fraction	0.7
selection_method	TOURNAMENT	fitness_prop_repro_fraction	0.1
function set	{+, −, *, %, sin, cos, exp}		

$$\text{hit} = \frac{N_{\text{up\_up}} + N_{\text{down\_down}}}{N_{\text{up\_up}} + N_{\text{up\_down}} + N_{\text{down\_up}} + N_{\text{down\_down}}}$$

$$= \frac{N_{\text{up\_up}} + N_{\text{down\_down}}}{30,177},$$

where $N_{\text{up\_up}}$ means the number of times when the tree makes an upward tendency while the observed price rises, and $N_{\text{down\_up}}$ means the number of times when the tree makes a downward tendency while the observed price falls, and so on. The total number of the predictions is 30,177, which equals the number of testing data.

All experimental results show that there seems to be a strong relationship among the MSE value, the hit percentage, and the profit gain. The lower the MSE value is, the higher both the hit percentage and the profit gain are. However, this is not necessarily a matter of course, because achieving the high profit requires more accurate prediction for the critical tendency change.

Table 3.22 shows that different weight values, i.e., w, resulted in different performance by STROGANOFF. We can observe that STROGANOFF gave relatively better performance under the condition A. The example acquired tree, i.e., the best evolved STROGANOFF predictors, under the condition A is shown in Fig. 3.16. The average and best hit percentages were well over 50% under the conditions A, B and D. Especially, STROGANOFF runs under the condition A resulted in the average hit percentage of 60% and over, which led to the high and positive profit gain. Using small weight values often gave rise to relatively long STROGANOFF trees so that the execution was aborted due to memory extinction. Fig. 3.19 shows the prediction of the normalized Nikkei225 price by the best evolved tree under the conditions A and E. The predicted value of Nikkei225 price for the first 100 minutes is shown for condition A. The predicted difference between the current Nikkei225 price and the price one minute before is plotted for condition E. Fig. 3.20 illustrates the optimal profit and the profits gained by the predicted trees. These results provide the evidence that the predicted difference under the condition E corresponds to the observed qualitative behavior, i.e., the upward

Table 3.21: Experimental results (STROGANOFF).

Condition	Weight	Training MSE	Testing Hit(%) Average	Testing Hit(%) Best	Testing Profit gain(yen) Average	Testing Profit gain(yen) Best
A	0.2	9.40e-06	62.3	62.4	30712	30762
	0.1	9.38e-06	62.3	62.4	30744	30762
	0.01	9.37e-06	62.2	62.3	30516	30823
	0.001	9.37e-06	62.2	62.4	30651	30804
	0.0001	9.37e-06	61.7	62.4	27511	30769
	0.0	9.38e-06	62.3	62.4	30654	30762
B	0.2	1.25e-05	57.5	57.7	18636	19194
	0.1	1.25e-05	57.3	57.7	18594	19194
	0.01	1.24e-05	55.3	57.7	13266	19194
C	0.2	6.57e-04	50.0	50.3	1599	3156
	0.1	6.57e-04	50.0	50.3	1517	3156
	0.01	6.57e-04	50.0	58.2	841	4044
	0.001	6.57e-04	49.9	50.1	890	1921
	0.0001	6.57e-04	50.0	50.8	1092	4044
	0.0	6.57e-04	50.0	50.2	471	2577
D	0.2	1.26e-05	57.6	57.7	18995	19194
	0.1	1.25e-05	57.2	57.7	18390	19194
	0.01	1.25e-05	54.9	57.7	13569	19194
E	0.2	7.25e-04	51.2	51.3	5785	6071
	0.1	7.24e-04	51.6	51.7	5381	5443
	0.01	7.24e-04	51.7	51.7	5443	5443
	0.001	7.24e-04	51.1	51.7	5381	5443
	0.0001	7.24e-04	51.7	51.7	5443	5443
	0.0	7.24e-04	51.7	51.7	5443	5443
	-0.01	7.24e-04	51.6	51.7	5381	5443

Table 3.22: Experimental results (GP).

Condition	Training MSE	Testing Hit(%) Average	Testing Hit(%) Best	Testing Profit gain(yen) Average	Testing Profit gain(yen) Best
A	1.79e-06	55.02	62.78	12411.01	31256.06
B	1.22e-05	47.47	48.17	−4093.22	−2341.50
C	5.82e-04	50.42	51.00	127.03	305.13
D	1.28e-05	41.09	51.64	−19727.52	−3811.19
E	1.80e-06	61.38	62.56	28942.03	30896.56

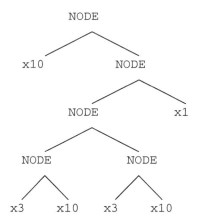

FIGURE 3.16: The best evolved tree by STROGANOFF under condition A.

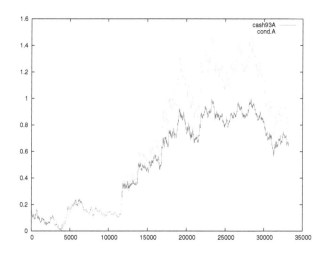

FIGURE 3.17: Time series predicted by STROGANOFF under condition A.

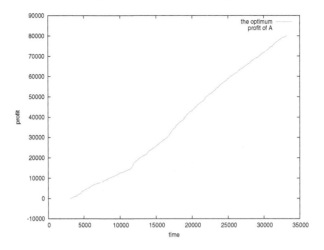

FIGURE 3.18: Profit gained by STROGANOFF under condition A.

or downward tendency, of the Nikkei225 price. This causes the high profit gain shown in Fig. 3.19.

Table 3.21 presents that the average and best hit percentages were below 50% by standard GP under the conditions B, C and D, which resulted in the low profit and the negative returns except the condition C. On the other hand, under the conditions A and E, the average hit percentage was over 50% and the best one was over 60%, which led to the high and positive profit gain. Especially, GP runs under the condition E resulted in the average hit percentage of 60% and over. Fig. 3.17 shows the prediction of the normalized Nikkei225 price by the best evolved tree under condition A. The predicted value (cond.A) of Nikkei225 price for the first 100 minutes is shown for condition A. The target Nikkei price (cash93A) is also shown in the figure. Fig. 3.17 illustrates the optimal profit and the profits gained by the predicted trees.

To summarize the above GP experimental results, we can confirm the following points:

1. The average or variance terminals were not effective for the prediction (conditions B and C).

2. Using only past data or difference values led to the unstable prediction (condition A).

3. The most effective terminal set included the absolute values and the directional values of the difference between the current Nikkei225 price and the past one (condition E).

Although the best profit is obtained by GP under condition A, the average profit is not necessarily high under the same condition. As can be seen

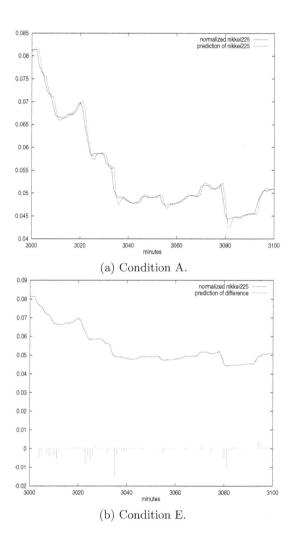

(a) Condition A.

(b) Condition E.

FIGURE 3.19: Prediction results by GP.

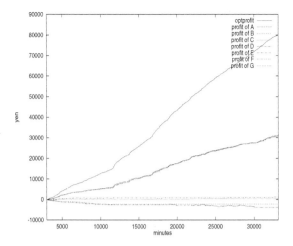

FIGURE 3.20: Optimal profit and profits gained by GP.

in these results, GP performance is extremely dependent upon the terminal choice. However, there is not much theoretical background for the best choice. In general, the terminal and function sets play an essential role in GP search, but they are problem-dependent and not easy to choose. On the other hand, STROGANOFF's performance is relatively stable independently from the terminal choice.

3.7.1.6 Comparative Experiment with Neural Networks

For the sake of comparison, we apply neural network to the same prediction task and examine the performance difference. We used the program available at "Neural Networks at Your Fingertips" [Kutza96]. This neural network program implements the classical multi-layer backpropagation network with bias terms and momentum. It is used to detect structure in time series, which is presented to the network using a simple tapped delay-line memory. The program originally learned to predict future sunspot activity from historical data collected over the past three centuries. To avoid over-fitting, the termination of the learning procedure is controlled by the so-called stopped training method.

The neural network parameters used are shown in Table 3.23. The network was trained under the previous condition A. That is, the input variables of the network were set to be $\{y1, \cdots, y10\}$. The random constant $\Re$ is omitted. Table 3.24 shows the experimental results. The data are averaged over 10 runs with different numbers of hidden units. Comparing these results with the ones in Tables 3.21 and 3.22, we can confirm that the neural network gave much worse results than STROGANOFF. The reason seems to be that the neural network suffers from the overfitting, as can be seen in the table.

Table 3.23: Neural network parameter.

#. of Layers	3	#. of hidden nodes	5, 10, 15
α	0.5	BIAS	1
η	0.05	EPOCHS	1000
Gain	1	#. of LOOP	100

Table 3.24: Experimental results (neural network).

	Training	Testing			
		Hit(%)		Profit gain(yen)	
#. hidden units	MSE	Average	Best	Average	Best
5	2.92e-06	58.2	60.2	23682	27586
10	2.70e-06	58.7	59.4	24725	26427
15	2.73e-06	58.3	59.5	23990	26245

Moreover, the computational time is much longer for the convergence for the neural network. Thus, we can conclude the superiority of STROGANOFF over the neural network.

3.7.1.7   Summary of the Results

With STROGANOFF, unlike in GP, historical values and the average terminal symbol worked effectively in forecasting.

In comparison of forecasting with STROGANOFF and simple GP, we have observed that the best values were nearly the same and that STROGANOFF was highly stable, even in the test runs and in no instance were the averages negative. Thus, STROGANOFF can be considered better suited to ordinary data than simple GP.

Although root mean square error in the training data in the STROGANOFF results was larger than simple GP, the best values for profit were nearly the same and the averages were high. The presumable reason is that as STROGANOFF was suited to more ordinary data, accuracy with respect to specific data decreased. Accordingly, presumably more ideal (ordinary) forecasting will be possible provided multiple sets of data from mutually different periods are used as the data used in training, several types of time series data (which are related) are used, and overall fitness to the data is evaluated. Many of the tree structures obtained in STROGANOFF implementation had shorter descriptions than those obtained in simple GP. This was possibly effective in the avoidance of over-training.

Forecasting using neural net is stable, it did not yield high profits such as those from GP and STROGANOFF. However, comparison by the number of intermediate layer units in neural net reveals that although error within the training data was nearly the same when the number was 10 or 15, error increased slightly when the number was 5. The best value for profit was obtained when the number of units was 5, and in time series data the results when the number of units was 10 or 15 were negatively affected by over-training.

3.7.2 Option Price Forecasting

3.7.2.1 Option Price

Option trading is the buying and selling of the rights to buy or sell certain goods. Some typical option transactions are shown in Table 3.25. An option that confers the right to sell is called a put option, and an option that confers the right to buy is called a call option. The price of the right in an option transaction is called the option price (premium). A call option involving share certificates is the right to buy shares at a certain price at a certain date and time. This "certain price" is called the strike price (or the exercise price).

For example, consider an option transaction in which someone has bought for 10,000JPY an option to buy one share of the stock of a certain company for 1,000,000JPY (a call option). The basis is the same in the case of a put option, except that the transaction is for the right to sell instead of the right to buy. If subsequently the share price on the maturity date were 1,100,000JPY and the person exercised the right and bought a share of the stock for 1,000,000JPY, the person could sell the share for 1,100,000JPY and earn a profit of 90,000JPY (100,000JPY minus the 10,000JPY price of the option). On the other hand, if the share price on the maturity date were 900,000JPY, the buyer would abandon the option and lose the 10,000JPY option price. Although in this transaction the loss for the buyer of the right and profit for the seller is at most 10,000JPY, the buyer's profit and seller's loss are theoretically infinite. Therefore, with option transactions it is possible for the buyer to limit loss. This is a highly attractive characteristic not found in stocks or futures. Fig. 3.21 shows stock price and profit for a certain transaction. The graph shows that provided the stock price exceeds the strike price (in this case, 1,000,000JPY), the buyer will exercise the right. Furthermore, in this transaction the margin of profit is higher than in the case of actual owner-ship of a share of the stock purchased at a price of 1,000,000JPY. That is, if a person earns a profit of 90,000JPY on the sale of a share purchased at a price of 1,000,000JPY, as the person's investment in the share is 1,000,000JPY, the margin of profit is 9%. However, if a person pays 10,000JPY as the pre-mium for an option and earns 90,000JPY in profit, the profit margin on the funds invested is 900%. Options such as this are highly leveraged products. Whereas many highly leveraged products are high-risk, high-return products such as future transactions offer the tremendous benefit of making it possible to limit loss. Option transactions can be used in hedging (avoiding the risk of future product price fluctuations) and investment and are likely to become increasingly important.

Table 3.25: Common option transactions in Japan.

	National government bond futures options
Tokyo Stock Exchange	Tokyo Stock Price Index (TOPIX) options
	Stock options
	Nikkei Stock Average (Nikkei225) options
Osaka Securities Exchange	Nikkei Stock Index 300 (Nikkei300) options
	Stock options
Nagoya Stock Exchange	Option 25 options
Tokyo International Financial Futures Exchange	Euro-yen three-month interest futures options

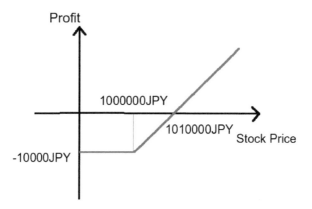

FIGURE 3.21: Stock price and profit.

Call options whose strike prices are lower than their respective current stock prices already have value. Such options are called in-the-money options (C in Fig. 3.22). Options whose strike prices and current stock prices are equivalent are called at-the-money options (B in Fig. 3.22). Options whose strike prices are higher than the current stock prices are called out-of-the-money options (A in Fig. 3.22). Out-of-the-money options have no value in relation to current share prices.

Option transactions are divided into two types according to when the buyer of the right can end the transaction. The first is the American type option, which the buyer can exercise at any time at the buyer's discretion. The second is the European type option, which cannot be exercised until the maturity date (delivery month). The Nikkei Stock Average options examined in this study are European type options.

As Nikkei Stock Average options' asset is the Nikkei Stock Average index, no delivery of instruments occurs, the rights for options that have value on

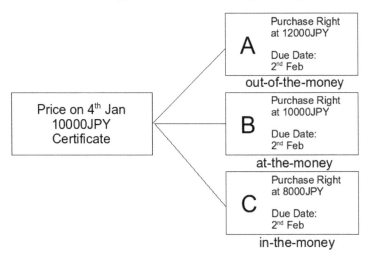

FIGURE 3.22: Option prices.

the maturity date are automatically exercised, and option holders receive the difference between the Nikkei Stock Average index and the strike price. Options that have no value on the maturity date are automatically abandoned. Nikkei Stock Average price option data is published on several web pages. These web pages provide a summary of the Nikkei Stock Average, the current option price value and the theoretical price for each call option and put option, strike prices, and other information.

3.7.2.2 The Black–Scholes Formula

The Black–Scholes formula is a famous method for calculating option prices (premiums), which is expressed as follows:

$$P = S \cdot N(\frac{u}{\sigma\sqrt{x}} + \sigma\sqrt{x}) - Xe^{-rx} \cdot N(\frac{u}{\sigma\sqrt{x}}), \tag{3.72}$$

$$u = \log\frac{S}{X} + (r - \frac{\sigma^2}{2})x, \tag{3.73}$$

where S represents the stock price, X represents the strike price, x represents the time to maturity of the option, δ represents volatility, r represents the risk-free interest rate and $N(d)$ represents the area from $-\infty$ to d under the standard normal distribution $N(0,1)$. $N(d)$ is calculated using the following formula (see Fig. 3.23):

$$N(d) = \int_{-\infty}^{d} \frac{1}{\sqrt{2\pi}} \exp(-\frac{z^2}{2}) dz. \tag{3.74}$$

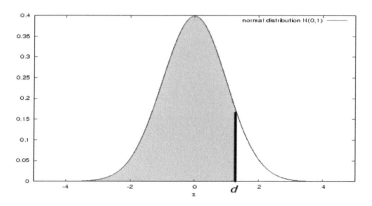

FIGURE 3.23: Standard normal distribution.

Furthermore, the formula for S in the Black–Scholes formula can be written as an (LISP) S-expression in the following way:

(-(∗ *S* normal(+(%(+ log om (∗(- rf (%(∗ hv hv) 2)) *x*)) (∗ hv *sqrt x*)) (∗ hv *sqrt x*))) (∗(% *x* exp(∗ rf *x*)) normal(%(+ log om (∗(- rf (%(∗ hv hv) 2)) *x*)) (∗ hv *sqrt x*))))

Although option prices can be calculated using the Black–Scholes formula, the theoretical prices calculated using this formula differ from actual market prices. The presumable reason for this is the various assumptions used to derive the Black–Scholes formula (that stock prices follow a Wiener process, the definition of portfolio, etc.). Accordingly, we will seek a more accurate option price equation using STROGANOFF.

3.7.2.3   Option Price Forecasting

As is the case with stock price forecasting, a great deal of research on option prices is conducted by using GP. For instance, [Chen *et al.*98] used actual market data – namely, the S&P 500 index (a stock price index calculated using the aggregate market value of 500 issues that represent key industries in the United States) options – and compared GP and the Black–Scholes formula. They used January 1995 as training data and February 1995 as test data, used symbols such as those shown in Table 3.26 as GP terminal symbols and non-terminal symbols, and aimed to calculate option prices using a common function. Also, [Chidambaran *et al.*98] used symbols such as those in Table 3.27 as terminal symbols and non-terminal symbols and applied GP to a virtual market generated using the Black–Scholes formula and a virtual market generated using a jump-diffusion process. In this research, a function that appears in the Black–Scholes formula is used as a function (N_{cdf} is the

Table 3.26: GP (non-)terminal symbols in [Chen *et al*.98].

Terminal set	τ	Time to maturity
	R_f	Risk-free interest rate (annual interest rate for U.S. government bonds)
	S/E	Stock price/strike price
	R	Random number
Functional set	$+, -, *, \%$	Four arithmetic functions
	sin	
	cos	
	exp	A maximum of 1,700
	log	

Table 3.27: GP (non-)terminal symbols in [Chidambaran *et al*.98].

Terminal set	S	Stock price
	X	Strike price
	S/X	Stock price/strike price
	τ	Time to maturity
	$(S - X)_+$	$\max(S - X, 0)$
Functional set	$+, -, *, \%$	Four arithmetic functions
	exp	
	log	
	$\sqrt{}$	Square root
	N_{cdf}	$N(d)$ in Fig. 3.23

above-mentioned $N(d)$ in Fig. 3.23). We referred to this research with respect to GP implementation conditions and other matters.

3.7.2.4 Data and Conditions

The data and conditions actually used in testing are as follows. In the test, we used Nikkei225 options data for June (624 samples) and July (508 samples) 2005. These are the only options in Japan that have sufficient liquidity. To avoid over-training and test the goodness of fit to ordinary data, we used the June data for training and the July data for the test. The acquired data were stock price, volatility, strike price, option price, and time to maturity of the option.

We used the six terminal symbols (i.e., inputs used for a neural network) of $S, X, S/X, \tau, rf, \sigma$, which are shown in Tables 3.26 and 3.27. For the sake of simplification, we used a constant risk-free interest rate (0.009).

We ran GP ten times using the parameters shown in Table 3.28, in which "sqrt" represents square root and "normal" represents the cumulative standard normal distribution function. We also ran STROGANOFF ten times for each

Table 3.28: GP parameters.

max_generation	100	max_depth_after_crossover	12
population_size	25,000	max_depth_for_new_trees	6
steady_state	0	max_mutant_depth	4
grow_method	GROW	crossover_any_pt_fraction	0.3
tournament_K	6	crossover_func_pt_fraction	0.5
selection_method	TOURNAMENT	fitness_prop_repro_fraction	0.1
function set	$+, -, \times, \div, \exp, \log, \mathrm{sqrt}, \mathrm{normal}$		

Table 3.29: STROGANOFF parameters.

population_size	5000
grow_method	GROW
crossover_func_pt_fraction	0.5
crossover_any_pt_fraction	0.399
WEIGHT	0.2, 0.15, 0.10, 0.05

weight using the parameters shown in Table 3.29 (parameters not included in the table are the same as for GP). In addition, for the sake of comparison, we ran neural network ten times for each number of units using the parameters shown in Table 3.30. As with stock price forecasting, to enable comparison of results for different numbers of units we used three values for the number of intermediate layer units: 5, 10, and 15.

3.7.2.5   Experimental Results

Table 3.31 shows the results of implementation under the above conditions. The values in the table indicate average error. "HP" is a comparison of web page theoretical values and actual values, and "BS" is a comparison of Black–Scholes formula values calculated by the authors and actual values.

Table 3.30: Neural network parameters.

Number of input layers	1
Number of hidden layers	1
Number of hidden units	5,10,15
Number of output layers	1
Number of input units	6
Number of output units	1
Number of training cycles	100
Number of Epoch	1000

Table 3.31: Comparative results.

Method	Average error training	test	Best result training	test
HP	26.9	54.1	–	–
BS	26.0	54.0	–	–
GP	24.3	34.1	19.0	20.8
NN(15)	19.7	28.8	18.5	22.9
NN(10)	20.0	29.3	19.0	26.8
NN(5)	24.6	32.9	19.1	25.8
STROGANOFF(0.2)	18.7	26.9	16.6	21.2
STROGANOFF(0.15)	17.9	27.1	16.5	23.9
STROGANOFF(0.1)	17.8	27.9	15.9	19.1
STROGANOFF(0.05)	17.0	26.5	16.0	17.6

The table shows that STROGANOFF produced the lowest average values and best values. Although forecasting using neural network was superior to GP in average values, GP was superior in best values. Also, all three methods (GP, STROGANOFF, neural network) yielded lower error values than the Black–Scholes formula.

Although there is some data spread in the GP and STROGANOFF results depending on the trial run, the best individual average error values were low for training and testing alike and better than the values for neural network. This result is attributable to the characteristics of the two techniques. As neural networks are local hill-climbing searches that use teacher signals to reduce error, in many cases they proceed to a certain point but stop at local solutions. By comparison, although GP and STROGANOFF global searches entail instability, in some cases they find extraordinarily good individuals.

However, GP entails difficulty in parameter selection: for instance, depending on the parameters, search efficiency decreases due to factors such as an increase in tree description length with the elapsing of generations. Table 3.32 shows the best individual formula obtained in GP. Here, rf represents risk-free interest rate, r represents time to maturity of the option, S represents stock price, X represents strike price, and om represents S/X.

In the neural network results, average values are midway between those of GP and STROGANOFF, and the best values were the worst of the three techniques. The results clearly show the characteristics of a local hill-climbing search. Also, although there was nearly no difference in values when the number of units in the option price forecasting formula was 10 or 15, the values were low when the number of units was 5. This confirms that, depending on the problem, in neural network adjustment of the number of intermediate layer units is important. In the results of STROGANOFF, the best values for training and testing were the best of the three techniques, and the expansion

Table 3.32: The best individual S formula obtained in a GP run.

(+(* (sqrt rf) (- τ (- (sin (* (sin (+ (* (* om (+ om (sin (exp (+
(* (exp om) om) (+ (* τ hv) om)))))) X) (* (sin om) (sin (- (- S
S) (+ τ (sqrt (* τ (exp (exp (+ rf om))))))))))))) om)) (+ (+ (* (+
(* τ hv) om) hv) om) (* (cos (sin (+ (* (* om (+ om (sin (exp
(+ (* (exp om) om) (exp om)))))) X) (* (sin om) (sin (- (- S S)
(+ τ (sqrt (* τ (exp (exp om))))))))))))) (+ (* (cos (plog (* om (*
(+ (* (exp (+ rf om)) om) (* τ X)) (* om (+ (* (exp om) om) (*
(+ (* om hv) om) om)))))))) (+ (* (* om (cos (* τ (ex (exp (+ (*
om hv) om)))))) X) (+ rf om))) rf))))))(* (sin om) om))

involving the addition of a statistical technique to a GP global search was a good fit. However, as STROGANOFF inherits the difficulty of parameter selection that is a drawback of GP, parameter adjustment is important. Table 3.33 shows the best individual formula (weight = 0.05). Comparison with the results of GP (Table 3.32) shows that although tree description length is shorter for STROGANOFF, the results for STROGANOFF are better. Accordingly, we were able to successfully suppress the tendency for GP tree description length to be too great, which presumably enabled an efficient search to proceed and favorable results to be obtained. Also, although the results in training data improved as we decreased the weight, a tendency toward over-training, which results in an increase in average error in test data over time, was found. The presumable reason for this is that having decreased the weight facilitated the selection of the trees with long descriptions specialized to the training data.

There remains room for improvement in option price forecasting in this research. One reason for error is that price data for periods in which nearly no trading occurred (too long until maturity) is included in the data used in the test. It is conceivable that excluding this data would further improve results. Also, modifications such as the division of option prices into the categories in-the-money and out-of-the-money before implementation can be considered. A very interesting outcome is the appearance of some differences from the normal distribution assumption used in the derivation of the Black–Scholes formula: individuals that did not use the normal distribution were found even among the individuals in GP, a technique for which the results were good; also, good results are obtained even in neural networks, STROGANOFF, and other polynomial approximation methods. Furthermore, although in this research we used data for a two-month period, it will be interesting to attempt to decide option prices in various circumstances by using data from other time periods.

Table 3.33: The best individual formula obtained in a STROGANOFF run.

Tree	NODE coeff.
(NODE291190960 (NODE291190992 om (NODE-291191024 (NODE291191056 (NODE135480176 τ om) (NODE291191088 (NODE291191120 (NODE291191152 (NODE291191184 (NODE-240994608 (NODE240994640 (NODE420833680 om S) (NODE135959264 om om)) om) (NODE-291191216 (NODE291191248 (NODE226626472 (NODE135959264 om om) (NODE135959264 om om)) τ) (NODE135480424 om S))) (NODE-160725328 τ om)) τ) (NODE135645712 S X))) om)) (NODE217261424 (NODE270195392 (NODE-153606944 (NODE331035864 τ (NODE301496448 S (NODE301496480 (NODE301496512 S om) τ))) (NODE182157864 (NODE335894720 (NODE-146551440 (NODE146551472 (NODE135479680 τ (NODE135479792 om om)) (NODE354759656 om om)) om) (NODE369693256 (NODE226626472 (NODE135959264 om om) (NODE135959264 om om)) X)) τ)) (NODE333471056 τ om)) (NODE-163123368 om (NODE163123400 τ S))))	NODE291190960: 1.751658e–03 1.652683e+00 –6.667349e–01 6.002723e–03 4.952423e–01 –4.846953e–01

3.7.2.6 Summary of the Results

First of all, the results for GP were slightly unstable, and tree descriptions tended to be long. This is presumably because, depending on the trial run, description portions that have no affect on overall operation increased during growth, tree description length increased even though nearly no change in fitness occurred, and search efficiency decreased. Accordingly, it is necessary to skillfully adjust the crossover parameters (the crossover_any_pt_fraction and crossover_func_pt_fraction in Table 3.28) for each problem. Also, as the results change significantly depending on modifications applied to the non-terminal symbols and terminal symbols, these settings are important as well.

Although with STROGANOFF we were able to achieve stable results with both types of data, as is the case with GP, it is necessary to experiment with adjustment of the terminal symbols for each problem and try various adjustments to the crossover parameters. Furthermore, it is also important to adjust weights and control tree growth.

With neural network, we were able to obtain stable, good results. Although the nature of a local hill-climbing technique is clearly evident in the results, best values are lower than for GP techniques and the results show proclivity to stop at local values. With regard to neural network, however, it is possible that changing the number of intermediate layers, the number of units, and other parameters will lead to the discovery of parameters appropriate to the problem.

3.8   Inductive Genetic Programming

Nikolaev and Iba have proposed inductive Genetic Programming (iGP) for the sake of extending STROGANOFF. This section describes the basics of iGP and its applications[3].

Inductive Genetic Programming is a specialization of the GP paradigm for inductive learning. The reasons for using this specialized term are: (1) *inductive* learning is a search problem and GP is a versatile framework for exploration of large multi-dimensional search spaces; (2) GP provides *genetic* learning operators for hypothetical model sampling that can be tailored to the data; and (3) GP manipulates *program*-like representations which adaptively satisfy the constraints of the task. An advantage of inductive GP is that it discovers not only the parameters but also the structure and size of the models.

The basic computational mechanisms of a GP system are inspired by those from natural evolution. GP conducts a search with a population of models using mutation, crossover, and reproduction operators. Like in the nature, these operators have a probabilistic character. The mutation and crossover operators choose at random the model elements that will undergo changes, while the reproduction selects random, good models among the population elite. Another characteristic of GP is its flexibility in the sense that it allows us easily to adjust its ingredients for the particular task. It enables us to change the representation, to tune the genetic operators, to synthesize proper fitness functions, and to apply different reproduction schemes.

3.8.1   Polynomial Neural Networks

Polynomial neural networks (PNNs) are a class of feedforward networks. They are developed with the intention of overcoming the computational limitations of the traditional statistical and numerical optimization tools for polynomial identification, which practically can only identify the coefficients of

---

[3]This section is mainly based on Nikolaev and Iba's recent works on the extension of STROGANOFF. The readers should refer to [Nikolaev and Iba06] for the details of iGP and other applications.

relatively low-order terms. The adaptive PNN algorithms are able to learn the weights of highly non-linear models.

A PNN consists of nodes, or neurons, linked by connections associated with numeric weights. Each node has a set of incoming connections from other nodes, and one (or more) outgoing connections to other nodes. All non-terminal nodes, including the fringe nodes connected to the inputs, are called hidden nodes. The input vector is propagated forward through the network. During the forward pass it is weighted by the connection strengths and filtered by the activation functions in the nodes, producing an output signal at the root. Thus, the PNN generates a non-linear real-valued mapping $P : \mathcal{R}^d \to \mathcal{R}$, which taken from the network representation is a *high-order polynomial model*:

$$P(\mathbf{x}) = a_0 + \sum_{i=1}^{L} a_i \prod_{j=1}^{d} x_j^{r_{ji}}, \qquad (3.75)$$

where a_i are the term coefficients, i ranges up to a preselected maximum number of terms L: $i \leq L$; x_j are the values of the independent variables arranged in an input vector $\mathbf{x}$, i.e., $j \leq d$ numbers; and $r_{ji} = 0, 1, \ldots$ are the powers with which the j-th element x_j participates in the i-th term. It is assumed that r_{ji} is bounded by a maximum polynomial order (degree) s: $\sum_{j=1}^{d} r_{ji} \leq s$ for every i. The above polynomial is linear in the coefficients a_i, $1 \leq i \leq L$, and non-linear in the variables x_j, $1 \leq j \leq d$.

Strictly speaking, a power series contains an infinite number of terms that can exactly represent a function. In practice a finite number of them are used for achieving the predefined sufficient accuracy. The polynomial size is manually fixed by a design decision.

3.8.2 PNN Approaches

The differences between the above PNN are in the representational and operational aspects of their search mechanisms for identification of the relevant terms from the power series expansion, including their weights and underlying structure. The main differences concern: (1) what is the polynomial network topology and especially what is its connectivity; (2) which activation polynomials are allocated in the network nodes for expressing the model, are they linear, quadratic, or highly non-linear mappings in one or several variables; (3) what is the weight learning technique; (4) whether there are designed algorithms that search for the adequate polynomial network structure; (5) what criteria for evaluation of the data fitting are taken for search control.

Tree-like genetic programs are suitable for iGP as they offer two advantages: (1) they have parsimonious topology with sparse connectivity between the nodes and (2) they enable efficient processing with classical algorithms. Subjects of particular interest here are the linear genetic program trees that are genotypic encodings of PNN phenotypes which exhibit certain input–output behaviors.

A genetic program has a *tree* structure. In it a node is below another node if the other node lies on the path from the root to this node. The nodes below a particular node are a subtree. Every node has a parent above it and children nodes under it. Nodes without children are leaves or terminals. The nodes that have children are non-terminals or functional nodes.

PNNs are represented with *binary trees* in which every internal functional node has a left child and a right child. A binary tree with Z functional nodes has $Z + 1$ terminals. The nodes are arranged in multiple levels, called also *layers*. The level of a particular node is one plus the level of its parent, assuming that the root level is zero. The *depth*, or height of a tree, is the maximal level among the levels of its nodes. A tree may be limited by a maximum tree depth, or by a maximum tree size which is the number of all nodes and leaves.

Trees are now described formally to facilitate their understanding. Let $\mathcal{V}$ be a vertex set from two kinds of components: *functional nodes* $\mathcal{F}$ and *terminal nodes* $\mathcal{T}$ ($\mathcal{V} = \mathcal{F} \cup \mathcal{T}$). A *genetic program* $\mathcal{G}$ is an ordered tree $s_0 \equiv \mathcal{G}$, in which the sons of each node $\mathcal{V}$ are ordered, with properties:

- it has a distinguishing parent $\rho(s_0) = \mathcal{V}_0$ called the *root* node;

- its nodes are labelled $\nu : \mathcal{V} \to \mathcal{N}$ from left to right and $\nu(\mathcal{V}_i) = i$;

- any functional node has a number of children, called arity $\kappa : \mathcal{V} \to \mathcal{N}$, and a terminal leaf $\rho(s_i) = \mathcal{T}_i$ has zero arity $\kappa(\mathcal{T}_i) = 0$;

- the children of a node $\mathcal{V}_i$, with arity $k = \kappa(\mathcal{V}_i)$, are roots of disjoint subtrees $s_{i1}, s_{i2}, \cdots, s_{ik}$. A subtree s_i has a root $\rho(s_i) = \mathcal{V}_i$, and subtrees $s_{i1}, \cdots, s_{ik}$ at its k children: $s_i = \{(\mathcal{V}_i, s_{i1}, s_{i2}, \cdots, s_{ik}) \mid k = \kappa(\mathcal{V}_i)\}$.

This labelling suggests that the subtrees below a node $\mathcal{V}_i$ are ordered from left to right as the leftmost child s_{i1} has the smallest label $\nu(s_{i1}) < \nu(s_{i2}) < \ldots < \nu(s_{ik})$. This ordering of the nodes is necessary for making efficient tree implementations, as well as for the design of proper genetic learning operators for manipulation of tree structures.

The construction of binary tree-like PNN requires us to instantiate its parameters. The terminal set includes the explanatory input variables $\mathcal{T} = \{x_1, x_2, ..., x_d\}$, where d is the input dimension. The function set contains the activation polynomials in the tree nodes $\mathcal{F} = \{p_1, p_2, ..., p_m\}$, where the number m of distinct functional nodes is given in advance. A reasonable choice is the incomplete bivariate polynomials up to second-order that can be derived from the complete one assuming that some of its coefficients are zero. The total number of such incomplete polynomials is 25 from all $2^5 - 1$ possible combinations of monomials $w_i h_i(x_i, x_j)$, $1 \leq i \leq 5$, having always the leading constant w_0, and two different variables. A subset $p_i \in \mathcal{F}$, $1 \leq i \leq 16$ of them is taken after elimination of the symmetric polynomials (Table 3.34).

Table 3.34: Activation polynomials [Nikolaev and Iba06, Table 2.1].

1.	$p_1(x_i, x_j) = w_0 + w_1 x_1 + w_2 x_2 + w_3 x_1 x_2$
2.	$p_2(x_i, x_j) = w_0 + w_1 x_1 + w_2 x_2$
3.	$p_3(x_i, x_j) = w_0 + w_1 x_1 + w_2 x_1 x_2$
4.	$p_4(x_i, x_j) = w_0 + w_1 x_1 + w_2 x_1 x_2 + w_3 x_1^2$
5.	$p_5(x_i, x_j) = w_0 + w_1 x_1 + w_2 x_2^2$
6.	$p_6(x_i, x_j) = w_0 + w_1 x_1 + w_2 x_2 + w_3 x_1^2$
7.	$p_7(x_i, x_j) = w_0 + w_1 x_1 + w_2 x_1^2 + w_3 x_2^2$
8.	$p_8(x_i, x_j) = w_0 + w_1 x_1^2 + w_2 x_2^2$
9.	$p_9(x_i, x_j) = w_0 + w_1 x_1 + w_2 x_2 + w_3 x_1 x_2 + w_4 x_1^2 + w_5 x_2^2$
10.	$p_{10}(x_i, x_j) = w_0 + w_1 x_1 + w_2 x_2 + w_3 x_1 x_2 + w_4 x_1^2$
11.	$p_{11}(x_i, x_j) = w_0 + w_1 x_1 + w_2 x_1 x_2 + w_3 x_1^2 + w_4 x_2^2$
12.	$p_{12}(x_i, x_j) = w_0 + w_1 x_1 x_2 + w_2 x_1^2 + w_3 x_2^2$
13.	$p_{13}(x_i, x_j) = w_0 + w_1 x_1 + w_2 x_1 x_2 + w_3 x_2^2$
14.	$p_{14}(x_i, x_j) = w_0 + w_1 x_1 + w_2 x_2 + w_3 x_1^2 + w_4 x_2^2$
15.	$p_{15}(x_i, x_j) = w_0 + w_1 x_1 x_2$
16.	$p_{16}(x_i, x_j) = w_0 + w_1 x_1 x_2 + w_2 x_1^2$

The notion of *activation polynomials* is considered in the context of PNN instead of transfer polynomials to emphasize that they are used to derive back-propagation network training algorithms.

The motivations for using all distinctive *complete* and *incomplete* (first-order and second-order) bivariate activation polynomials in the network nodes are: (1) having a set of polynomials enables better identification of the interactions between the input variables; (2) when composed higher-order polynomials rapidly increase the order of the overall model, which causes over-fitting even with small trees; (3) first-order and second-order polynomials are fast to process; and (4) they define a search space of reasonable dimensionality for the GP to explore. The problem of using only the complete second-order bivariate polynomial is that the weights of the superfluous terms do not become zero after least mean square fitting, which is an obstacle for achieving good generalization.

The following expression illustrates a hierarchically composed polynomial extracted from the PNN in Fig. 3.24 to demonstrate the transparency and easy interpretability of the obtained model.

```
(( w0 + w1 * z7^2 + w2 * z4^2 )
     z7=( w0 + w1 * x2 + w2 * x2^2 + w3 * x3^2 )
        x2
        x3 )
     z4=( w0 + w1 * z2 + w2 * z2 * x1 + w3 * z2^2 )
```

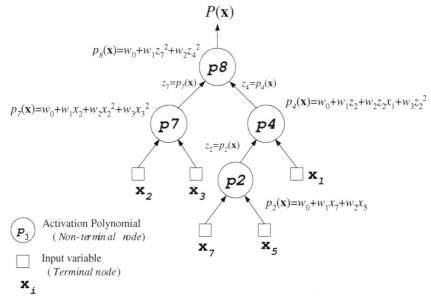

FIGURE 3.24: Tree-structured representation of a PNN [Nikolaev and Iba06, Fig. 2.1].

```
z2=( w0 + w1 * x7 + w2 * x5 )
      x7
      x5 )
   x1 ))
```

The accommodation of a set of complete and incomplete activation polynomials in the network nodes makes the models versatile for adaptive search, while keeping the neural network architecture relatively compact. Using a set of activation polynomials does not increase the computational cost for performing GP. The benefit of having a set of activation polynomials is of enhancing the expressive power of this kind of PNN representation.

An example of a tree-structured polynomial using some of these activation polynomials is illustrated in Fig. 3.24. The computed polynomial $P(\mathbf{x})$ at the output tree root is the multivariate composition: $P(x_1, x_2, x_3, x_5, x_7) = p_8(p_7(x_2, x_3), p_4(p_2(x_7, x_5), x_1))$.

3.8.3 Basic iGP Framework

The iGP paradigm can be used for the automatic programming of polynomials. It provides a problem independent framework for discovering the polynomial structure, in the sense of shape and size, as well as the weights. The iGP learning cycle involves five substeps: (1) ranking of the individuals according to their fitness; (2) selection of some elite individuals to mate

Table 3.35: Basic framework for iGP [Nikolaev and Iba06, p.49].

Inductive Genetic Programming

Step	Algorithmic sequence
1. Initialization	Let the generation index be $\tau = 0$, and the population size be n
	Let the initial population be: $\mathcal{P}(\tau) = [g_1(\tau), g_2(\tau), ..., g_n(\tau)]$ where g_i, $1 \leq i \leq n$, are genetic programs of depth up to S
	Let μ be a mutation parameter, κ be a crossover parameter
	Create a random initial population: $\mathcal{P}(\tau) = RandomTrees(n)$, such that $\forall g, Depth(g) < S$
	Evaluate the fitnesses of the individuals: $F(\tau) = Evaluate(\mathcal{P}(\tau), \lambda)$ and order the population according to $F(\tau)$
2. Evolutionary Learning	a) Select randomly $n/2$ elite parents from $\mathcal{P}(\tau)$ $\mathcal{P}'(\tau) = Select(\mathcal{P}(\tau), F(\tau), n/2)$
	b) Perform recombination of $\mathcal{P}'(\tau)$ to produce $n/4$ offspring $\mathcal{P}''(\tau) = CrossTrees(\mathcal{P}'(\tau), \kappa)$
	c) Perform mutation of $\mathcal{P}'(\tau)$ to produce $n/4$ offspring $\mathcal{P}''(\tau) = MutateTrees(\mathcal{P}'(\tau), \mu)$
	d) Compute the offspring fitnesses $F''(\tau) = Evaluate(\mathcal{P}''(\tau), \lambda)$
	e) Exchange the worst $n/2$ from $\mathcal{P}(\tau)$ with offspring $\mathcal{P}''(\tau)$ $\mathcal{P}(\tau + 1) = Replace(\mathcal{P}(\tau), \mathcal{P}''(\tau), n/2)$
	f) Rank the population according to $F(\tau + 1)$ $g_0(\tau + 1) \leq g_1(\tau + 1) \leq \cdots \leq g_n(\tau + 1)$
	g) Repeat the Evolutionary Learning with another cycle $\tau = \tau + 1$ until the termination condition is satisfied

and produce offspring; (3) processing of the chosen parent individuals by the crossover and mutation operators; (4) evaluation of the fitness of the offspring; and (5) replacement of predetermined individuals in the population by the newly born offspring. Table 3.35 presents the basic iGP algorithmic framework.

The formalization of the basic framework, which can be used for implementing an iGP system, requires some preliminary definitions. The iGP mechanisms operate at the genotype level – that is they manipulate linearly implemented genetic program trees g. The basic control loop breeds a population $\mathcal{P}$ of genetic programs g during a number of cycles τ called generations. Let n denote the size of the population vector – that is the population includes $g_i, 1 \leq i \leq n$ individuals. Each individual g is restricted by a predefined tree depth S and size L in order to limit the search space to within reasonable bounds. The initial population $\mathcal{P}(0)$ is randomly created.

The function *Evaluate* estimates the fitness of the genetic programs using the fitness function f to map genotypes $g \in \Gamma$ into real values $f \colon \Gamma \to R$. The fitness function f takes a genetic program tree g, decodes a phenotypic PNN model from it, and measures its accuracy with respect to the given data. All the fitness values of the genetic programs from the population are kept in an array F of size n. The selection mechanism *Select* $\colon \Gamma^n \to \Gamma^{n/2}$ operates according to a predefined scheme for picking randomly $n/2$ elite individuals which are going to be transformed by crossover and/or mutation.

The recombination function *CrossTrees* $\colon \Gamma^{n/4} \times R \to \Gamma^{n/4}$ takes the half $n/4$ from the selected $n/2$ elite genetic programs, and produces the same number of offspring using size-biased crossover using parameter κ. The mutation function *MutateTrees* $\colon \Gamma \times R \to \Gamma$ processes half $n/4$ from the selected $n/2$ elite genetic programs, using size-biased context-preserving mutation using parameter μ.

The resulted offspring are evaluated, and replace inferior individuals in the population *Replace* $\colon \Gamma^{n/2} \times \Gamma^{n/2} \times N \to \Gamma^n$. The steady-state reproduction scheme is used to replace the genetic programs having the worst fitness with the offspring so as to maintain a proper balance of promising individuals. Next, all the individuals in the updated population are ordered according to their fitness values.

The readers should refer to [Nikolaev and Iba06] for more details of iGP process and its application results.

3.9   Discussion

3.9.1   Comparison of STROGANOFF and Traditional GP

The previous sections showed the experimental results of our system STRO-GANOFF. This section discusses the effectiveness of our numerical approach to GP.

Due to the difficulties mentioned in Section 3.2.2, we have observed the following inefficiencies with traditional GP:

1. The number of individuals to be processed for a solution is much greater than with other methods.

2. Over-fitting occurs in time series prediction tasks (Table 3.4).

3. Randomly generated constants do not necessarily contribute to the desired tree construction, because there is no tuning mechanism for them.

To overcome these difficulties, we have introduced a new approach to GP, based on a numerical technique, which integrates a GP-based adaptive search of tree structures, and a local parameter tuning mechanism employing statistical search. Our approach has overcome the GP difficulties mentioned in Section 3.2.2 in the following ways:

1. GP search is effectively supplemented with the tuning of node coefficients by multiple regression. Moreover, STROGANOFF can guide GP recombination effectively in the sense that the recombination operation is guided using MDL values (Section 3.3.8).

2. MDL-based fitness evaluation works well for tree structures in STRO-GANOFF algorithm, which controls GP-based tree search.

3. STROGANOFF performance is less affected by the terminal choice than GP's (Tables 3.21 and 3.22).

First, node coefficients can be tuned by our statistical method. This tuning is done "locally", in the sense that coefficients of a certain data point are derived from the data of its child nodes. Thus, STROGANOFF integrates the local search of node tuning with GP-based global search. Furthermore, as described in Section 3.3.8, this mechanism together with MDL values leads to the recombinative guidance of STROGANOFF.

Second, MDL-based fitness is well-defined and used in our STROGANOFF trees. This is because a STROGANOFF tree has the following features:

Size-based Performance The more the tree grows, the better its performance (fitness) is. This is a basis for evaluating the trade-off between the tree description and the error.

Decomposition The fitness of a substructure is well-defined, i.e., the fitness of a subtree (substructure) reflects that of the whole structure. If a tree has good substructures, its fitness is high.

The complexity-based fitness evaluation has already been introduced in order to control GA search strategies. We have shown that an MDL-based fitness can also be used for controlling the tree growth in STROGANOFF, i.e., an MDL-based fitness prevents over-generalization in learning. The effectiveness of an MDL-based fitness definition for GP has also been discussed in [Iba and Sato94b] and [Zhang and Mühlenbein95].

Third, as we have observed financial applications (Tables 3.21 and 3.22), STROGANOFF performs less dependently upon the terminal choice than GP. This feature is desirable in the sense that the best choice of terminals is not always known beforehand. Also note that although the best profit is obtained by GP under condition A, the average profit of GP is not necessarily high

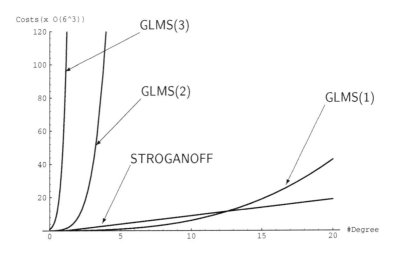

FIGURE 3.25: Computational costs [Iba *et al.*96a, Fig. 9].

under the same condition. Thus, we can believe that the STROGANOFF's performance is more stable than GP, which is more suitable for the real-world applications.

3.9.2   Computation Costs

The calculation of the inverse matrix (equation (3.17)) consumes much less computation time, because STROGANOFF requires the coefficients for only two variables at each non-terminal node. Fig. 3.25 plots the computation costs with the degrees of fitting polynomials. The costs are estimated as the numbers of the loop iterations of the inverse calculations. GLMS(i) represents the general least mean square method for the fitting equation of i input variables. The vertical axis is translated (i.e., divided by $O(6^3)$) for the sake of convenience. As can be seen in the figure, the advantage of STROGANOFF comes about when dealing with large, complex systems, i.e., when fitting a higher-order polynomial of multiple input variables as explained below.

Because $X'X$ of equation (3.17) is a matrix, whose size is the number of terms of a fitting equation (e.g., equations (3.13)~(3.15)), the multiple regression analysis requires the inverse calculation of a matrix of such size. Now let us consider the number of terms of a fitting equation for the general linear least square method. The number of terms in a complete multinomial of degree n (i.e., the sum of all homogeneous multinomials from 0-th degree through n-th degree) in m variables is given as follows [Farlow84]:

$$N_C(n, m) = \frac{(n + m)!}{n! \times m!}. \tag{3.76}$$

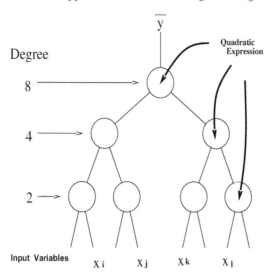

FIGURE 3.26: A GMDH Tree [Iba *et al.*96a, Fig. 13].

Computing an inverse of $K \times K$ matrix requires $O(K^3)$ loop executions by means of either Gaussian elimination or LU decomposition [Press *et al.*88, p.38]. Therefore, the computational cost of the general least mean square method for m input variables is given as,

$$O(N_C^3) = O(\{\frac{(n+m)!}{n! \times m!}\}^3). \tag{3.77}$$

These costs are plotted as GLMS(m) in Fig. 3.25. As can be seen in the figure, finding coefficients by this method is clearly out of question for multiple input variables.

On the other hand, the GMDH process in STROGANOFF is able to find the higher-order regression polynomial by repeatedly solving two-variable regressions of low-order. If we use the following quadratic expression,

$$z(x_1, x_2) = a_0 + a_1 x_1 + a_2 x_2 + a_3 x_1 x_2 + a_4 x_1^2 + a_5 x_2^2, \tag{3.78}$$

the computational cost for the inverse matrix is estimated as $O(6^3)$. Fig. 3.26 shows the repetition of these multiple regressions for a GMDH tree. As you can see, it is necessary to construct a d-depth binary tree for getting a 2^d-degree expression. A d-depth binary tree contains $2^d - 1$ internal nodes. Therefore, the number of inverse matrix calculations is $2^d - 1$ in order to obtain a multiple regression of a 2^d-degree expression with 2^d input variables. In other words, the computational cost for a GMDH tree for an N-degree regression is given as:

$$(N - 1) \times O(6^3). \tag{3.79}$$

This computational cost is plotted in Fig. 3.25 (i.e., STROGANOFF). The figure shows the advantage of STROGANOFF over the general least mean square method, especially in the case of regression of a multiple-input higher-order equation.

To conclude, the STROGANOFF (or its GMDH tree) is superior in terms of computational costs for large, complex systems.

3.9.3 Genetic Programming with Local Hill-Climbing

The main feature of our work is that our approach introduces a way to modify trees, by integrating node coefficient tuning and traditional GP recombination. Our numerical approach builds a bridge from traditional GP to a more powerful search strategy. We have introduced a new approach to GP by supplementing it with a local hill-climbing approach. Local hill-climbing search uses local parameter tuning (of the node functionality) of tree structures, and works by discovering useful substructures in STROGANOFF trees. Our proposed augmented GP paradigm can be considered schematically in several ways:

augmented GP	=	**global search**	+	**local hill-climbing search**
	=	structured search	+	parameter tuning of node functionalities

The local hill-climbing mechanism uses a type of relabelling procedure[4], which finds a locally (if not globally) optimal assignment of nodes for an arbitrary tree. Therefore, speaking generally, our new approach can be characterized as:

augmented GP	=	traditional GP	+	relabelling procedure

The augmented GP algorithm is described below:

Step1 Initialize a population of tree expressions.

Step2 Evaluate each expression in the population.

Step3 Create new expressions (children) by mating current expressions. Apply mutation and crossover to the parent tree expressions.

Step4 Replace the members of the population with the child trees.

Step5 A local hill-climbing mechanism (called "relabelling") is executed periodically, so as to relabel nodes of the trees of the population.

Step6 If the termination criterion is satisfied, then halt; else go to **Step2**.

---

[4]The term "label" is used to represent the information (such as a function or polynomial) at a non-terminal node.

Table 3.36: Properties of STROGANOFF variants [Iba *et al*.96a, Table 1].

	BF-STROGANOFF	**STROGANOFF**	**ℜ-STROGANOFF**
Problem domain	Boolean concept formation	System identification	Temporal data processing
Tree type	Binary tree		Network
Terminal nodes	Input variables, their negations	Input variables	Input variables
Functional nodes	AND, OR, LEFT, RIGHT	Polynomial relationships	Polynomial relation-ships, memory
Relabelling process	ALN	GMDH	Error-propagation
Reference	[Iba and Sato94b]	[Iba *et al*.93] [Iba *et al*.96a]	[Iba *et al*.95]

As can be seen, **Steps**1–5 follow traditional GP, where **Step**4 is the new local hill-climbing procedure. In our augmented GP paradigm, the traditional GP representation (i.e., the terminal and non-terminal nodes of tree expressions) is constrained so that our new relabelling procedure can be applied. The sufficient condition for this applicability is that the designed representation has the property of "insensitivity" or "semantic robustness", i.e., changing a node of a tree does not affect the semantics of the tree. In other words, the GP representation is determined by the choice of the local hill-climbing mechanism.

In this chapter, we have chosen a GMDH algorithm as the relabelling procedure for system identification problems. We employed other relabelling procedures for various kinds of problem domains. For instance, in [Iba *et al*.94a] and [Iba *et al*.95], we have chosen two vehicles to perform the relabelling procedure, known respectively as ALN (Adaptive Logic Network [Armstrong91]) and the error propagation. The characteristics of these resulting GP variants are summarized in Table 3.36.

This process is totally called as Memetic Algorithms (MA), which is one of the recent growing areas of research in evolutionary computation (see [Lim09] for details).

3.9.4   Applicability to Computational Finances

The previous experimental results have shown the effectiveness of STRO-GANOFF approach for the sake of predicting financial data. However, there are several points to be improved for practical use. For instance, the following extensions should be considered:

1. The dealing simulation should be more realistic including the payment of the commission. The profit gain is offset with the fee.

2. The prediction error should be improved. Especially, we should put much more emphasis on the short-term or real-time prediction, rather than the long-term prediction.

3. The problem-specific knowledge, such as economical index options or foreign exchange rates, could be introduced for the further performance improvement.

The method of buying and selling shares used in our experiment is not very realistic. In particular, the use of available funds for only one stock at a time and the failure to consider commissions are problems. Accordingly, we will attempt to expand the test to handle multiple stocks and take commissions into consideration.

As for the third point, we are now in pursuit of the quantitative factor analysis for the purpose of choosing the significant economical features. This will have an essential impact on the prediction accuracy, especially for the short-term prediction.

STROGANOFF is a numerical GP system, which effectively integrates traditional GP adaptive search and statistical search [Iba *et al.*96a]. The preliminary results obtained by STROGANOFF were satisfactory and promising. However, we also observed the overfitting difficulty. This is probably because STROGANOFF used the polynomial regression, which led to finding the highly fit polynomials in terms of MSE or MDL values. But this did not necessarily give rise to the high profit gain as mentioned earlier. We believe that this difficulty will be avoided by using the discrete terminals, such as a step function or a sign function.

3.9.5 Limitations and Further Extensions

Whereas traditional GP relies upon a large population to maintain diversity, and requires only several generations, our method can function with a small population, and can construct useful building blocks as the generations proceed. Also, the total number of evaluations of individuals is probably much less for STROGANOFF. For instance, we showed that the computational effort of STROGANOFF was 20 to 50 times less than that of traditional GP, for several symbolic regression problems [Iba *et al.*96a], and that the number of individuals to be processed by traditional GP (for the same quality of solution) in the time series prediction problem was much greater than that of STROGANOFF (Table 3.4). However this difference does not reflect the difference in computational complexities between the two, because a STROGANOFF evaluation involves many regression derivations. Most of the computational burden concentrates on the multiple regression analysis (i.e., the derivation of the inverse matrix, equation(3.17)). We have not yet studied the computational complexity of STROGANOFF theoretically. Thus it is difficult to compare the proposed algorithm with other approaches. The purpose of this chapter is to propose a numerical approach to GP and to show its feasibility through experiment. Theoretical studies, including a mathematical analysis of the computational complexities of STROGANOFF, and the improvement of its efficiency, remain important research topics.

One limitation of our approach is the memory space required for statistical calculation. In general, each intermediate node requires the storage of a set of data, whose size is equal to that of the training data. For instance, consider the P_1 tree in Fig. 3.4. Let N be the number of training data. In order to derive the coefficients $(b_0, \cdots, b_5)$ of NODE2 (z_2), N data of (z_1, x_3) are used to deduce N equations of (3.14). Thus N values of z_1 should be kept in NODE3 rather than be calculated on request, for the purpose of saving the computation of the same z_1 values for later usage. Therefore a large memory space may be needed for the entire population of GMDH trees in our STROGANOFF system. Another limitation is the computational time needed to perform the multiple regression analysis, as mentioned above. However, we believe that parallelizing STROGANOFF (i.e., both the GP process and the statistical process) leads to a reduction of the computational cost.

3.10 Summary

This chapter introduced a numerical approach to GP, which integrates a GP-based adaptive search of tree structures and a statistical search technique. We have established an adaptive system called STROGANOFF, whose aim is to supplement traditional GP with a local parameter-tuning mechanism. More precisely, we have augmented the traditional structural search of GP with a local hill-climbing search which employs a relabelling procedure. The effectiveness of this approach to GP has been demonstrated by its successful application to numerical and symbolic problems.

In addition, we described a new GP-based approach to temporal data processing, and presented an adaptive system called $\Re$-STROGANOFF. The basic idea was derived from our previous system STROGANOFF. $\Re$-STROGANOFF integrates an error-propagation method and a GP-based search strategy. The effectiveness of our approach was confirmed by successful application to an oscillation task, to inducing languages from examples, and to extracting finite-state automata (FSA).

We have also applied STROGANOFF to such "real world" problems as predicting stock-market data or developing effective dealing rules. We presented the application of STROGANOFF to the prediction of stock-price data in order to gain the high profit in the market simulation. We confirmed the following points empirically:

1. STROGANOFF was successfully applied to predicting the stock-price

data. That is, the MSE value for the training data was satisfactorily low, which gave rise to the high profit gain in the dealing simulation.

2. The performance under a variety of conditions, i.e., different terminal sets, was compared. Using the terminals based upon the delayed difference of the stock-price were more effective than using the exact price values.

3. The STROGANOFF result was compared with those of neural networks and GP, which showed the superiority of our method.

As for the future, we intend to extend STROGANOFF by:

- parallelization

- performing a theoretical analysis of computational complexities

Another important area of research concerns the extension of STROGANOFF framework to other symbolic applications, such as concept formation or program generation. We believe the results shown in this chapter are a first step towards this end.

Chapter 4

Classification by Ensemble of Genetic Programming Rules

This chapter presents various feature selection methods, classification methods, fitness evaluation methods, and ensemble techniques. Since the success of a genetic algorithm or genetic programming depends on a number of factors, including the fitness function used to evaluate the fitness of an individual, we present various fitness evaluation techniques for classification of balanced and unbalanced data. In many classification problems, single rules of genetic programming do not produce good test accuracy. To improve the performance of genetic programming, the majority voting genetic programming classifier (MVGPC) has been presented. The effectiveness of MVGPC is demonstrated by applying it to the classification of microarray data and unbalanced data. Though adaptive boosting (AdaBoost) is widely used as an accuracy improvement ensemble technique, we show that MVGPC performs better than AdaBoost for certain types of problems, such as classification of microarray data and unbalanced data. Finally, for an ensemble technique, various performance improvement techniques, such as diversity measurement and reduction of execution time, are discussed in detail.

4.1 Background

4.1.1 Data Classification Problem

Data classification is the problem of categorization of an item given the quantitative information of the various features that describe the item. In this context, first, an item is defined in terms of various characteristics, and then data about a set of items of known categories are acquired from various sources. The unknown category of an item is determined with a classifier that is built using this set of items of known categories.

Formally, the problem can be stated as follows: given a set of N training items $\{(\mathbf{x}_1, y_1), (\mathbf{x}_2, y_2), \ldots, (\mathbf{x}_N, y_N)\}$, build a hypothetical classifier $h : \mathbf{X} \to \mathbf{Y}$ that maps an item $\mathbf{x} \in \mathbf{X}$ to a class label $y \in \mathbf{Y}$.

Let us give an example of the classification problem. Suppose that we want

Table 4.1: An example of training data used to build a classifier.

Item	Features				Class
	a	b	c	d	label
x_1	13	87	134	59	Bad
x_2	15	76	116	59	Bad
x_3	13	81	131	52	Bad
x_4	12	64	100	58	Good
x_5	10	83	145	53	Bad
x_6	14	50	123	56	Good
x_7	16	81	142	54	Bad
x_8	16	51	130	55	Bad
x_9	19	99	102	53	Good
x_{10}	13	60	103	57	Good
x_{11}	17	67	138	51	Bad
x_{12}	17	90	102	51	Good
x_{13}	12	76	112	50	Good
x_{14}	18	96	110	59	Bad
x_{15}	20	94	134	57	Bad

to know whether an item is good or bad given the numerical values of the four features. For this purpose, first, we build a training data set of 15 items; the items are shown in Table 4.1. Using this training data, a classifier is learned, which is as follows:

$$Class(\mathbf{x}) = \begin{cases} \text{Good if } ((a < 15 \text{ AND } b < 75)\text{OR}(c < 125 \text{ AND } d < 55)); \\ \text{Bad} \quad \text{otherwise} \end{cases}$$

where $\mathbf{x}$ is a vector of values of the four features a, b, c, and d. In this chapter, we use the term *instance* to mean an item.

After a classifier is built from the training data, its performance is evaluated on the independent test data (also called validation data) to determine how accurately the classifier will classify the unknown instances. Sometimes, the training and test data are available; sometimes, only training data are available. When only training data are available, the performance of a classifier is evaluated through a technique called cross-validation. Moreover, during the selection of important features from data, the cross-validation technique is widely used for the evaluation of the goodness score of a feature subset. In k-fold cross-validation, the data D is randomly partitioned into k mutually exclusive subsets, $D_1, D_2, \ldots, D_k$ of approximately equal size. The classifier is trained and tested k times; each time $i(i = 1, 2, \ldots, k)$, it is trained with $D \backslash D_i$ and tested on D_i. The sequence of steps in the cross-validation technique is given in Fig. 4.1. After cross-validation, we get the summary of classification statistics, namely, the numbers of true positives (N_{TP}), true negatives (N_{TN}),

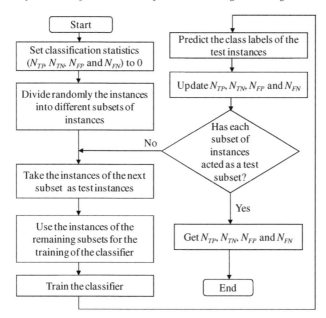

FIGURE 4.1: The sequence of steps in a cross-validation technique.

false positives (N_{FP}), and false negatives (N_{FN}). These statistics are usually converted to a single value, such as accuracy or F-score. When k is equal to the number of instances in the data set, it is called leave-one-out cross-validation (LOOCV) [Kohavi95]. The cross-validation accuracy is the overall number of correctly classified instances, divided by the number of instances in the data. When a classifier is stable for a given data set under k-fold cross-validation, the variance of the estimated accuracy would be approximately equal to $\frac{accuracy(1-accuracy)}{N}$ [Kohavi95], where N is the number of instances in the data set. When the number of instances in a data set is very small, usually LOOCV is used. An example of the LOOCV technique is given in Fig. 4.2.

4.1.2 Feature Selection Problem

Sometimes, not all the features of the data are important for the classification of the instances, and in certain types of data, especially in DNA microarray data, there are many redundant features. These irrelevant and redundant features sometimes have a negative effect on the accuracy of the classifier and increase data acquisition cost as well as learning time. Moreover, a large number of features hide the interpretability of the learning model. Feature selection, also known as attribute selection or dimensionality reduction, is a technique for selecting a subset of relevant features from data, which is widely

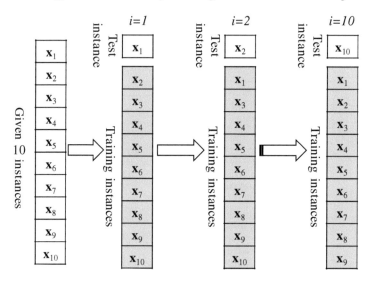

FIGURE 4.2: An example of the leave-one-out cross-validation technique.

used in artificial intelligence and machine learning to build a fast and robust classifier. Here, an example of feature selection is given. For the training data in Table 4.2, a classifier that can perfectly classify all the training instances is given as:

$$Class(\mathbf{x}) = a \text{ XOR } b \text{ XOR } c \text{ XOR } e$$

where XOR operation is defined as follows:

$$a \text{ XOR } b = (a \text{ AND NOT}(b)) \text{ OR } (\text{NOT}(a) \text{ AND } b).$$

In respect of this classifier, feature f is redundant as $a = f$, and feature d is irrelevant and has a negative effect on the classifier. By removing these two features from the data, we get a subset of four important features, a, b, c, and e, that build a perfect classifier.

For a given classifier and a training data set, an optimal subset of features can be found by exhaustively searching all possible subsets of features. For a data set with n features, there are 2^n feature subsets. So, it is impractical to search the entire space exhaustively, unless n is small. There are two approaches, filter and wrapper approaches [Kohavi and John97], for selection of features. The two approaches are illustrated in Figs. 4.3 and 4.4.

In the filter approach, the data are preprocessed, and some top-rank features are selected using a quality metric, independently of the classifier. Though the filter approach is computationally more efficient than the wrapper approach, it ignores the effects of the selected features on the performance of the classifier but the selection of the optimal feature subset is always dependent on the classifier.

Table 4.2: An example data set with irrelevant and redundant features.

Instance	Features						Class
	a	b	c	d	e	f	label
x_1	1	1	1	1	0	1	1
x_2	0	1	1	1	1	0	1
x_3	1	0	0	1	0	1	1
x_4	1	1	1	0	1	1	0
x_5	1	1	0	1	1	1	1
x_6	0	1	0	0	1	0	0
x_7	0	0	0	1	0	0	0
x_8	1	0	0	0	1	1	0
x_9	1	1	0	0	0	1	0
x_{10}	1	1	1	1	0	1	1
x_{11}	1	0	0	1	0	1	1
x_{12}	0	1	0	0	1	0	0
x_{13}	1	0	1	0	0	1	0
x_{14}	0	0	0	1	0	0	0
x_{15}	0	0	0	1	0	0	0

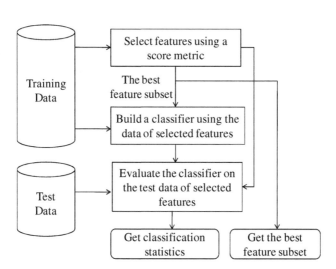

FIGURE 4.3: Filter approach of selection of features, and evaluation of the selected features on test data.

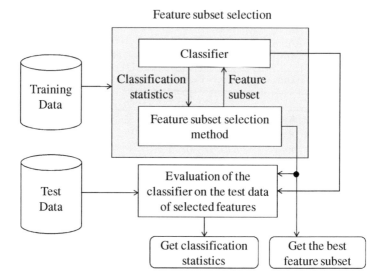

FIGURE 4.4: Wrapper approach of selection of features, and evaluation of the selected features on test data.

In the wrapper approach, the feature subset selection algorithm conducts the search for a good subset by using the classifier itself as a part of the evaluation function. The classification algorithm is run on the training data, partitioned into internal training and validation data, with various feature subsets. The internal training data are used to estimate the parameters of a classifier, and the internal validation data are used to estimate the goodness of a feature subset. The feature subset with the highest estimated fitness is chosen as the final set on which the classifier is run. Usually in the final step, the classifier is built using the whole training data and the final feature subset, and then evaluated on the test data. The major disadvantages of the wrapper approach are that it requires much computation time and the selected features are classifier-specific.

4.2 Various Classifiers

Classifiers are used to predict the class of an unknown instance. Class prediction is a supervised learning method and various machine-learning classifiers, such as support vector machine (SVM)[Vapnik95], C4.5 (decision tree) [Quinlan93], k-nearest neighbor (k-NN)[Dasarathy91], weighted voting classifier (WVC)[Golub *et al.*99, Slonim *et al.*00], and naive Bayes classifier (NBC), to name a few prominent classifiers, are used as class predictors. In this chapter, weighted voting classifier, naive Bayes classifier, and k-nearest neighbor classifier are described in detail.

4.2.1 Weighted Voting Classifier

A classifier based on weighted voting has been proposed in [Golub *et al.*99, Slonim *et al.*00]. We use the term *weighted voting classifier* to refer to this classifier. To determine the class of an instance, a weighted voting scheme has been used. The vote of each feature is weighted by the correlation of that feature with a particular class. The weight of a feature X_i is the signal-to-noise ratio defined as

$$W(X_i) = \frac{\mu_1 - \mu_2}{\sigma_1 + \sigma_2} \tag{4.1}$$

where μ_1, σ_1 and μ_2, σ_2 are the mean and standard deviation of the values of feature X_i in class 1 and 2, respectively. (Note that $W(X_i) = SNR(X_i)$.) The weighted vote of a feature X_i for an unknown instance $\mathbf{x} = (x_1, x_2, \ldots, x_n)$ is

$$V(X_i) = W(X_i) \left(x_i - \frac{\mu_1 + \mu_2}{2} \right) \tag{4.2}$$

where x_i is the value of feature X_i in that unknown instance $\mathbf{x}$. Then, the class of the instance $\mathbf{x}$ is

$$class(\mathbf{x}) = sign \left\{ \sum_{X_i \in G} V(X_i) \right\} \tag{4.3}$$

where G is the set of selected features. If the computed value is positive, the instance $\mathbf{x}$ belongs to class 1; negative value means $\mathbf{x}$ belongs to class 2. This classifier is applicable to binary classification problems.

4.2.1.1 Prediction Strength of WVC

It is always preferable for a classifier to give a confidence measure (prediction strength) of a decision about the class of a test instance. By defining a minimum confidence level for classification, one can decrease the number of false positive and false negatives at the expense of increasing the number of unclassified instances. The combination of a good metric for decision confidence and a good threshold value of that metric will result in a low false positive and/or low false negative rate without a concomitant high unclassified sample rate [Keller *et al.*00]. The choice of appropriate decision confidence metric depends on the particular classifier and how the classifier is employed.

The authors in [Golub *et al.*99] and [Slonim *et al.*00] defined the prediction strength (ps) for the weighted voting classifier as follows:

$$ps = \left| \frac{V_+ - V_-}{V_+ + V_-} \right| \tag{4.4}$$

where V_+ and V_- are respectively the absolute values of sum of all positive $V(X_i)$'s and negative $V(X_i)$'s calculated using equation (4.2).

The classification of an unknown instance is accepted if $ps > \theta$ (θ is the prefixed prediction strength threshold); otherwise the instance is classified as undetermined. Sometimes, undetermined instances are treated as misclassified instances.

4.2.2   Naive Bayes Classifier

The naive Bayes classifier (NBC) uses a probabilistic approach to predict the class of an instance. It computes the conditional probabilities of various classes given the values of the features and predicts the class with the highest conditional probability. During calculation of conditional probability, it assumes the conditional independence of features.

Let C denote a class from the set of m classes, $\{c_1, c_2, \ldots, c_m\}$, $\mathbf{X}$ is an instance described by a vector of n features, i.e., $\mathbf{X} = (X_1, X_2, \ldots, X_n)$; the values of the features are denoted by the vector $\mathbf{x} = (x_1, x_2, \ldots, x_n)$. The naive Bayes classifier tries to compute the conditional probability $P(C = c_i | \mathbf{X} = \mathbf{x})$ (or in short $P(c_i|\mathbf{x})$) for all c_i's and predicts the class for which this probability is the highest. Using Bayes' rule, we get

$$P(c_i|\mathbf{x}) = \frac{P(\mathbf{x}|c_i)P(c_i)}{P(\mathbf{x})} . \tag{4.5}$$

Since NBC assumes the conditional independence of features, equation (4.5) can be written as

$$P(c_i|\mathbf{x}) = \frac{P(x_1|c_i)P(x_2|c_i)\cdots P(x_n|c_i)P(c_i)}{P(x_1, x_2, \ldots, x_n)} . \tag{4.6}$$

The denominator in (4.6) can be neglected, since for a given instance, it is fixed and has no influence on the ranking of classes. Thus, the final conditional probability takes the following form:

$$P(c_i|\mathbf{x}) \propto P(x_1|c_i)P(x_2|c_i)\cdots P(x_n|c_i)P(c_i) . \tag{4.7}$$

Taking logarithm we get,

$$\ln P(c_i|\mathbf{x}) \propto \ln P(x_1|c_i) + \cdots + \ln P(x_n|c_i) + \ln P(c_i) . \tag{4.8}$$

For a symbolic (nominal) feature,

$$P(x_j|c_i) = \frac{\#(X_j = x_j, C = c_i)}{\#(C = c_i)} \tag{4.9}$$

where $\#(X_j = x_j, C = c_i)$ is the number of instances that belong to class c_i, and feature X_j has the value of x_j, and $\#(C = c_i)$ is the number of instances that belong to class c_i. If a feature value does not occur given some classes, its conditional probability is set to $\frac{1}{2N}$, where N is the number of instances. For a continuous feature, the conditional probability density is defined as

$$P(x_j|c_i) = \frac{1}{\sqrt{2\pi}\sigma_{ji}} e^{-\frac{(x_j-\mu_{ji})^2}{2\sigma_{ji}^2}} \qquad (4.10)$$

where μ_{ji} and σ_{ji} are the expected value and standard deviation of feature X_j in class c_i. Taking logarithm of (4.10) we get,

$$\ln P(x_j|c_i) = -\frac{1}{2}\ln(2\pi) - \ln \sigma_{ji} - \frac{1}{2}\left(\frac{x_j - \mu_{ji}}{\sigma_{ji}}\right)^2. \qquad (4.11)$$

Since the first term in (4.11) is constant, it can be neglected during calculation of $\ln P(c_i|\mathbf{x})$.

The advantage of the naive Bayes classifier is that it is simple and can be applied to multiclass classification problems.

4.2.2.1 Prediction Strength of NBC

For the naive Bayes classifier, the prediction strength metric for binary classification problems can be defined as the relative log likelihood difference of the winner class from the loser class [Keller *et al*.00]. That is, the prediction strength of the classifier for an unknown instance $\mathbf{x}$ is

$$ps = \frac{\ln P(c_{winner}|\mathbf{x}) - \ln P(c_{loser}|\mathbf{x})}{\ln P(c_{winner}|\mathbf{x}) + \ln P(c_{loser}|\mathbf{x})}. \qquad (4.12)$$

4.2.3 k-Nearest Neighbor Classifier

The k-nearest neighbor (k-NN) classifier [Dasarathy91], an extension of nearest neighbor classifier (IB1)[Aha *et al*.91], has long been used in pattern recognition and machine learning for supervised classification tasks. The basic approach involves storing all the training instances; when a test instance is presented, retrieving k training instances that are nearest to this test instance and prediction of the label of the test instance by majority voting. The distance between two instances $\mathbf{x} = (x_1, x_2, \ldots, x_n)$ and $\mathbf{y} = (y_1, y_2, \ldots, y_n)$ is calculated as follows:

$$d(\mathbf{x}, \mathbf{y}) = \sqrt{\sum_{i=1}^{n} w_i(x_i - y_i)^2} \qquad (4.13)$$

where n is the number of features in the data set and w_i is the weight of feature i. When $w_i = 1$ is set to 1, the distance between two instances becomes Euclidean distance. To avoid a tie, the value of k should be an

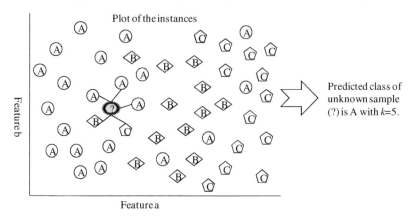

FIGURE 4.5: An example of class prediction by the k-NN classifier.

odd number (and of course, $k <$ number_of_training_samples) for binary classification. An example of the k-NN classifier is given in Fig. 4.5.

k-NN is very easy to implement and can provide good classification accuracy if the features are chosen and weighted carefully in computation of distance.

When the instances are of variable lengths or discrete time series data, such as the health checkup and lifestyle data of an organization, distance between two instances cannot be calculated by directly applying the above equation (4.13). In [Paul *et al.*08a], two methods are proposed, namely, the aggregation method and the sliding window method, to calculate the distance between two variable length instances and determine the nearest neighbors.

In the aggregation method, the length of each instance is made equal by aggregating the values of a feature across various observations into a single value, and then distance between two equal length instances is calculated. The aggregation method that is to be applied for a feature depends on the data type of that feature. If a feature is numeric, a function that returns a numeric value should be used; if the feature is nominal, a function that returns a nominal value should be used. Some examples of aggregation functions for numeric features are *average*, *max*, and *min*; one example of an aggregation function for nominal features is *mode* that returns the most frequent value.

In the sliding window method, distances between two instances are calculated at different window positions of the larger instance, and the window having the minimum distance is taken as the best matching window. The window is moved one data-point at a time where each data-point consists of either the value of a single observation, such as the output signal amplitude of a sensor at time t or the values of a number of features under single observation, such as the health checkup and lifestyle data of a person in a year. If the number of data-points in two instances is respectively m and n ($n <= m$), the number of sliding windows to consider to find the minimum distance is

$m - n + 1$. Let us give an example of the sliding window method to find the minimum distance in two instances consisting of the following data-points: $X_1 X_2 X_3 X_4 X_5$ and $Y_1 Y_2 Y_3$. To find the minimum distance, three sliding windows are considered and distances between $X_1 X_2 X_3$ and $Y_1 Y_2 Y_3$; $X_2 X_3 X_4$ and $Y_1 Y_2 Y_3$; and $X_3 X_4 X_5$ and $Y_1 Y_2 Y_3$ are calculated.

4.3   Various Feature Selection Methods

In this section, some feature selection methods are described in detail.

4.3.1   Ranking of Features

Widely used score metrics for ranking of features are signal-to-noise ratio (SNR) [Golub *et al.*99], *t*-statistics [Ding03], threshold number of misclassification score [Ben-Dor *et al.*00], likelihood score [Keller *et al.*00], and disorder score [Park *et al.*01].

4.3.1.1   Signal-to-Noise Ratio

The traditional rank-based feature selection method selects those features that individually best classify the training instances. A widely used method for evaluation of how well a feature separates the instances is the signal-to-noise ratio [Golub *et al.*99]. The SNR of a feature X_i in binary classification is defined as

$$SNR(X_i) = \frac{\mu_1 - \mu_2}{\sigma_1 + \sigma_2} \qquad (4.14)$$

where μ_1 and σ_1, and μ_2 and σ_2 are means and standard deviations of the values of feature X_i in class 1 and 2, respectively. Features with the most positive and the most negative $SNR(X_i)$ values are selected in parallel and are grouped together in equal numbers in the final classifier. For a multiclass problem having c classes, $\frac{c(c-1)}{2}$ pairwise classifiers are considered. If we need to select n features in total, we select $\frac{2n}{c(c-1)}$ distinct features from each pair i and j. For each pair, we apply the above rule of binary classification to select top-rank features. If some of the selected features of each pair are already included in the final subset, we exclude those features and select the features next to them in the ranking list. For binary classification, if n is odd, we select $\lceil \frac{n}{2} \rceil$ features having the most positive values, and $\lfloor \frac{n}{2} \rfloor$ features having the most negative values.

Variant forms of the signal-to-noise ratio have also been used for feature ranking. For example, the authors in [Furey *et al.*00] use the absolute value of $SNR(X_i)$ as the ranking criterion, and the authors in [Pavlidis *et al.*01] use $\frac{(\mu_1)^2 - (\mu_2)^2}{(\sigma_1)^2 + (\sigma_2)^2}$ as the score metric for feature ranking.

The serious limitation of this method of feature selection is that it may include some redundant features and exclude those complementary features that individually do not separate data well.

4.3.1.2 *t*-Statistics

In [Ding03], the author uses *t*-test to select important features from microarray data.

The *t*-test score of a feature X_i is defined as follows:

$$t\text{-value} = \frac{\mu_1 - \mu_2}{\sigma}; \sigma^2 = \frac{(n_1 - 1)\sigma_1^2 + (n_2 - 1)\sigma_2^2}{n - 2} \tag{4.15}$$

where μ_1 and σ_1, and μ_2 and σ_2 are means and standard deviations of the values of feature X_i in class 1 and 2, respectively; n_1 and n_2 are the numbers of instances in class 1 and 2, respectively. Then, the features having higher absolute *t*-values are selected as the discriminative features.

4.3.2 Sequential Forward Selection

Sequential forward selection (SFS) (also called forward sequential selection) [Aha and Bankert95] is a greedy, deterministic heuristic search algorithm that starts from an empty set of features and sequentially adds the feature that results in the highest evaluation value, such as the classification accuracy, when combined with the features already selected. It continues until no improvement is achieved in the evaluation value. For a given classifier, if the optimal subset of features contains a very small number of features, SFS performs the best because it performs the major part of its search near the empty feature set. The major disadvantage of SFS is that some of its previously selected features may become irrelevant or act negatively after addition of some strongly correlated features. Let us give an example of this drawback of SFS. Suppose that a data set has five features: a, b, c, d, and e. When used individually, they obtain classification accuracy of 45%, 60%, 35%, 15%, and 20%, respectively. Therefore, in the first step, b will be selected. Next suppose that the combination of (b, a),(b, c), (b, d), and (b, e) produces 70%, 62%, 65%, and 67% accuracy, respectively. In the second step, a is selected as it obtains the highest classification accuracy when combined with b. Suppose that in the third step, d is selected as the combination (a, b, d) obtains 80% accuracy. However, if the combination (a, d) is able to produce 80% accuracy, feature b becomes irrelevant. In the third step, even if the combination (a, b, d) obtains 75% accuracy, d will be selected but b in that case acts negatively on the acquired accuracy of the combination (a, d).

4.3.3   Genetic Algorithm-Based Methods

Genetic algorithms (GAs)[Goldberg89, Holland75], based on natural selection and adaptation, are developed to solve complex real-world problems. GAs are widely used to solve those problems that are highly non-linear, contain inaccurate and noisy data, and whose objective function cannot be expressed mathematically. Recently, genetic algorithms and their variants (parallel genetic algorithm, multi-objective evolutionary algorithm, and probabilistic model building genetic algorithm) have been applied to selection of informative features from microarray data [Liu and Iba01, Liu and Iba02, Deb and Reddy03, Ooi and Tan03, Deutsch03, Ando and Iba04, Li *et al.*04, Kim *et al.*04, Liu *et al.*05]. These methods usually employ a wrapper approach [Kohavi and John97] of feature selection, where a classifier is used to measure the goodness of a feature subset. These methods obtain better classification accuracies than ranking-based feature selection methods because various combinations of features are evaluated in evolutionary computation through generation of various individuals of a population. However, the success of identification of a smaller-size predictive feature subset depends on the choice of the appropriate recombination operators of an evolutionary computation method as well as on the choice of the appropriate classifier. We use the terms *an individual* and *a feature subset* interchangeably to refer to a subset of features. The sequence of steps in a typical GA-based method is as follows:

```
Procedure GeneSelectionByGA
 BEGIN
  Set the values of various controlling parameters;
  Generate initial population randomly;
  Evaluate each individual in the population;
  WHILE termination_criteria NOT satisfied DO
   BEGIN
    Select some top-ranked individuals from the population
     of previous generation;
    Generate offspring by applying crossover and mutation
     on the selected individuals;
    Evaluate offspring;
    Generate new population by combining offspring
       and old population;
   END
  Return the best individual from the population;
 END
```

First, the values of various controlling parameters, such as the population size (P), offspring size (O), the maximum number of generations, crossover probability and mutation probability, are set. In a GA, each candidate solution is evaluated using some score metrics, and some of the better solutions are selected for reproduction. There are two main genetic operators, crossover and mutation, for reproduction of offspring. Crossover operator creates offspring by combining parts from two or more parents (selected solutions); whereas, mutation generates an offspring by making changes in a single solution. The offspring are evaluated, and some of them are combined with the old population to generate the new population. This completes one cycle of generation. After several generations, the algorithm terminates converging to the optimal or a suboptimal solution.

The main points to consider during application of a genetic algorithm are:

- how to encode an individual;

- how to generate the initial population;

- how to evaluate the fitness of an individual;

- which genetic operators to apply to generate offspring;

- how to select individuals for genetic operations;

- when to terminate the program; and

- how to control various parameters, such as population size, offspring size, the maximum number of generations, crossover probability, and mutation probability.

4.3.3.1   Encoding of an Individual

In selection of informative features from data, an individual (feature subset) of a genetic algorithm population is encoded as a binary string with one bit for each feature. If a bit is "1", it means that the feature is selected in the feature subset; "0" means its absence. For example, in the binary string "1000100111," features 1, 5, 8, 9, and 10 of a data set are selected as possible informative features. The main advantage of these binary-coded individuals is that the design of a crossover or a mutation operator is easy.

4.3.3.2   Initial Population Generation

Individuals of the initial population for the problem are usually generated by random compositions of 1's and 0's. In this type of initialization, half of the features of the data set are expected to be selected in each individual. When the number of features in a data set is large, such as in microarray data, sometimes it becomes difficult to obtain compact feature subsets through the applications of recombination operators, and the initial best fitness does not improve over generations. To get compact feature subsets, the number of selected features in each individual of the initial population can be restricted to be an integer in $[1, m]$ where $m << n$ and n is the number of features in the data set.

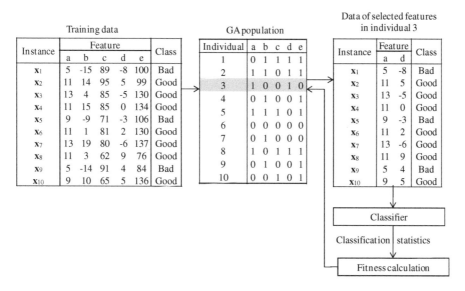

FIGURE 4.6: Fitness evaluation in a genetic algorithm.

4.3.3.3 Evaluation of an Individual

In Fig. 4.6, we have shown the steps in calculation of fitness of a feature subset. First the data corresponding to the selected features are extracted from the training data. Then the extracted data are passed to a classifier that classifies the data using a cross-validation method and returns the classification statistics, namely, the numbers of true positives (N_{TP}), true negatives (N_{TN}), false positives (N_{FP}), and false negatives (N_{FN}), which are then utilized by the fitness calculation method. Various fitness evaluation methods are described later in this chapter.

4.3.3.4 Selection of Individuals for Generation of Offspring

In the next step of a genetic algorithm, a number of individuals are selected based on fitness for reproduction of offspring using crossover and mutation. This artificial selection process imitates Darwinian natural selection—survival for the fittest; that is, the fitter an individual is, the more chances it gets to reproduce. Many selection methods have been proposed in the field of evolutionary computation; some of them are: roulette-wheel selection, rank-based selection, truncation selection, and tournament selection. Refer to Chapter 2 for a detailed description of various selection methods.

4.3.3.5 Generation of Offspring

In a typical genetic algorithm, offspring are generated by applying crossover and mutation to the selected individuals. Since the problem is binary-coded, one-point, two-point, or multi-point crossover and bit mutation are widely used. In one-point crossover, the individuals are cut at some randomly chosen point and the resulting sub-portions are swapped. In two-point crossover, the individuals are cut at two random points and the resulting sub-portions are swapped. In uniform crossover, one offspring is generated by randomly selecting a bit from either of the parents. For example, if $< 0111011011 >$ and $< 1010101010 >$ are two selected individuals, one possible offspring could be $< 0010111011 >$.

4.3.3.6 Creation of New Population

The old population and the newly generated feature subsets are combined to create the new population. There are different strategies, such as elitism and CHC selection [Eshelman91], to create the new population. The CHC stands for cross generational elitist selection, heterogeneous recombination (by incest prevention) and cataclysmic mutation. In elitism, $(P > O)$, and the top $(P - O)$ feature subsets of the old population are retained, and the remaining feature subsets are replaced with the newly generated ones. In CHC, $O = P$, and the new population is generated by selecting the best P feature subsets from the combination of $(P + O)$ feature subsets.

4.3.3.7 Termination Criteria

The termination criteria are user-defined. Widely used termination criterion is that the maximum number of generations has passed, the best feature subset in the population has reached the optimum fitness or the fitness of the best individual in the population does not improve in consecutive generations. After the termination of the algorithm, the selected features in the best individual in the population are taken as the important features of the target problem.

4.3.4 RPMBGA+

RPMBGA+ [Paul *et al.*08b, Paul *et al.*08a] belongs to the group of algorithms called probabilistic model building genetic algorithms (PMBGAs) [Pelikan *et al.*99a] (also called estimation of distribution algorithms (EDAs) [Mühlenbein and Paaß96]) and is an extension of the random probabilistic model building genetic algorithm (RPMBGA) [Paul and Iba05]. Detailed descriptions of various estimation of distribution algorithms can be found in [Larrañaga and Lozano01, Paul and Iba03a, Paul and Iba03b]. RPMBGA+ is a global search heuristic like genetic algorithm but instead of using the crossover and mutation operators of genetic algorithms, it generates offspring by sampling the probabilities of the features being selected in an individual. For this purpose, RPMBGA+ maintains a vector of probability of a feature being selected in an individual in addition to the population of individuals. Each individual in RPMBGA+ is a vector of 0's and 1's where "0" means that feature is not selected and "1" means that feature is selected. Each value $P(X_i, t) \in [0, 1](i = 1, 2, \ldots, n)$ in the vector of probability $P(\mathbf{X}, t)$ indicates the probability of the feature i being selected in a candidate feature subset in generation t. The pseudocode of RPMBGA+ is given below.

```
Procedure RPMBGA+
BEGIN
    Set the values of various controlling parameters;
    t=0;
    Initialize probability vector P(X,t);
    Generate initial population by sampling P(X,t);
    Evaluate each individual in the population;
    WHILE Termination_Criteria NOT satisfied DO
      BEGIN
        Select some top-ranked individuals from the population;
        Calculate marginal distributions M(X,t) of the features;
        Update probability vector P(X,t+1);
        Generate offspring by sampling P(X,t+1);
        Evaluate each offspring;
        Create new population using old population and offspring;
        t=t+1;
      END
    Return the best individual from the population;
  END
```

The first step in RPMBGA+ is to initialize the vector of probabilities and set the values of various controlling parameters, such as population size, offspring size, and the maximum number of generations. Usually, the initial probability of a feature being selected as important is set to 0.5, which means that the feature may or may not be selected as an important or relevant feature depending on the training data. However, in some cases, some information about the relationship of a feature with the target class is known. For example, it is known that salty and oily food cause high blood pressure. In RPMBGA+, this known information about the relationship of a feature with the target class is utilized during the initialization of the probability vector and the update of the probability vector. In RPMBGA+, the probability of each feature is initialized in the following way:

$$P(X_i, 0) = \begin{cases} p_i & \text{if } p_i > 0; \\ 0.5 & \text{otherwise} \end{cases} \tag{4.16}$$

where p_i is the prior information about the relationship of the feature i with the target class. When no information is known about the relationship of the feature with the target class, the probability is set to 0.5.

Given a probability vector $P(\mathbf{X}, t)$, the value $X_i (i = 1, 2, \ldots, n)$ in an individual corresponding to the feature i is generated in the following way:

$$X_i = \begin{cases} 1 & \text{if } P(X_i, t) \geq r_i; \\ 0 & \text{otherwise} \end{cases} \tag{4.17}$$

where $r_i \in [0, 1]$ is a random number that is generated by calling the random number generator in a programming language. Let us give an example of creation of an individual for a data set consisting of five features. Suppose that the probability vector is $(0.75, 0.92, 0.15, 0.29, 0.41)$ and the randomly generated five numbers are $(0.84, 0.39, 0.43, 0.27, 0.56)$. Therefore, the generated individual will be 01010.

Then the individuals in the population are evaluated using a classifier and a fitness function. If the termination criterion is not met, the probability vector is updated by using the probabilities of the features in the previous generation and their probability distributions calculated from the selected individuals. Usually in probabilistic model building genetic algorithms, the truncation selection is applied to select individuals; some top-ranked individuals are picked from the population for update of the probability vector. Different PMBGAs have applied different strategies for update of the probability vector. For example, in population-based incremental learning (PBIL) [Baluja94], $P(X_i, t + 1)$ is updated in the following way:

$$P(X_i, t + 1) = \alpha P(X_i, t) + (1 - \alpha) M(X_i, t) \tag{4.18}$$

where $\alpha \in [0, 1]$ is called the learning rate and fixed through each iteration, and $M(X_i, t)$ is the marginal probability distribution of X_i in the selected individuals. In RPMBGA, $P(X_i, t + 1)$ is updated as follows:

$$P(X_i, t + 1) = \alpha \beta P(X_i, t) + (1 - \alpha)(1 - \beta) M(X_i, t) \tag{4.19}$$

where $\alpha \in [0, 1]$ is called the learning rate and fixed through each iteration, and $\beta \in [0, 1]$ is a random number and changes at each iteration. Owing to the inclusion of an extra random parameter less than or equal to 1.0, $P(X_i, t + 1)_{RPMBGA} < P(X_i, t + 1)_{PBIL}$. Therefore, when the number of features in a data set is huge, RPMBGA will return a smaller-size feature subset than PBIL. However, after a number of generations, the probability becomes so small that most of the individuals are empty feature subsets. To prevent this, the lower limit of $P(X_i, t + 1)$ is set to a very small value, for example, to 0.001, and the value depends on the number of features in the data.

In RPMBGA+, the known information about the relationship of a feature with the target class is utilized during update of the probability vector in the following way:

$$P(X_i, t + 1) = \begin{cases} p_i & \text{if } p_i > 0; \\ \alpha \beta P(X_i, t) + (1 - \alpha)(1 - \beta) M(X_i, t) & \text{otherwise} \end{cases} \tag{4.20}$$

where p_i is the prior information about the relationship of the feature i with the target class.

Next, offspring are generated by sampling the probability vector and evaluated in the same way as initial individuals. Similar to a genetic algorithm, a new population is created from the old population and the offspring by applying a regenerational strategy, such as elitism or CHC. An example of the flow of steps in a generation of RPMBGA+ starting from the initial population with no prior information is shown in Fig. 4.7. After the termination criteria are met, the selected features in the best individual of the population are returned as the most important features.

4.4   Classification by Genetic Programming

For selection of important features, we need a feature selection method as well as a classifier, and the selected features and the accuracy of classification are highly dependent on the choice of the classifier. For example, for some data sets, support vector machine and genetic algorithm produces the better results, whereas for some other data sets, the k-NN classifier with RPMBGA produces the better results. The optimal tuning of all the parameters in the feature selection method and in the classifier is more difficult than that in a single method. Instead of two methods, genetic programming [Koza92, Banzhaf *et al.*98] can be used for both tasks. In its typical implementation, a set of classification rules is produced in multiple runs using the training data, and then the best fitted rule(s) is (are) used as the predicted model that is evaluated on the test data. The features in the best rule(s) are taken as the most important features. The number of rules needed for a classification problem depends on the number of classes of instances in the data; for binary classification, one rule is sufficient but for multiple classification, we need multiple rules.

Similar to a GA, there are a number of issues concerning the application of genetic programming to the classification of instances: encoding of a classification rule, generation of the initial population, evaluation of a classification rule, generation of offspring through crossover and mutation, and generation of multiple rules for multiclass classifications, etc. Some of these issues are described here in the context of data classification with illustrations and examples. Refer to Chapter 2 for description of other issues, such as generation of initial population, and generation of offspring through crossover and mutation. The flow of steps in a typical GP for classification is shown in Fig. 4.8.

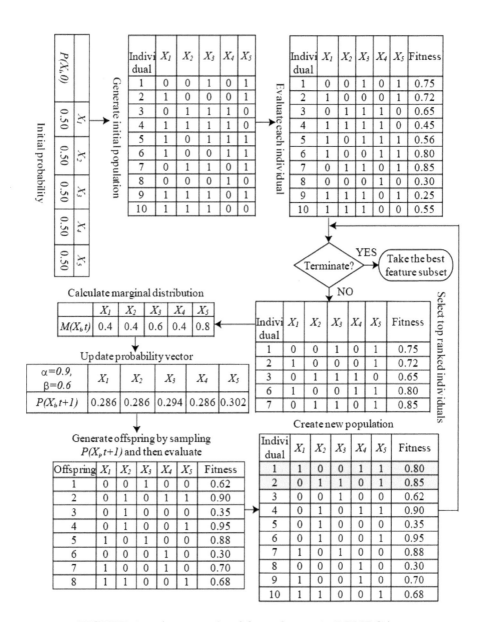

FIGURE 4.7: An example of flow of steps in RPMBGA+.

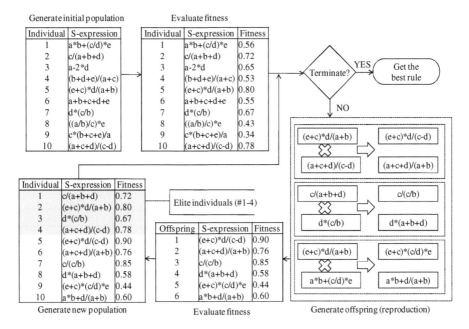

FIGURE 4.8: Flow of steps in a typical GP.

4.4.1   Encoding of a Classification Rule

Encoding is a technique of mapping a solution space into a search space. A good encoding can reduce the search space, whereas a bad encoding forces the algorithm to explore a huge search space. Usually, in genetic programming an individual is represented as a tree structure that can be easily evaluated in a recursive manner. Each tree may consist of a single rule or multiple rules, such as decision trees. A single rule tree usually represents an S-expression of terminals and functions like $(X1 - X3 + X5/X6)$ or $((X1 = X2)AND(X5 <> X9)$ where X_i corresponds to the i-th feature in the data set, and is widely used in binary classification [Paul and Iba07, Yu *et al.*06b, Paul *et al.*06, Mitra *et al.*06, Muni *et al.*04, Tan *et al.*02, Hong and Cho04, Chien *et al.*02, Falco *et al.*02, Zhang and Smart06, Langdon and Buxton04, Tackett93]. Every tree node has a function and every terminal node has one operand, making mathematical expressions easy to evolve and evaluate. By using this S-expression, the class of an instance is predicted in the following way:

- Output of the S-expression is a continuous value:

    IF (S-expression ≥ 0) THEN TargetClass ELSE OtherClass.

- Output of the S-expression is a binary value (true or false):

    IF (S-expression) THEN TargetClass ELSE OtherClass.

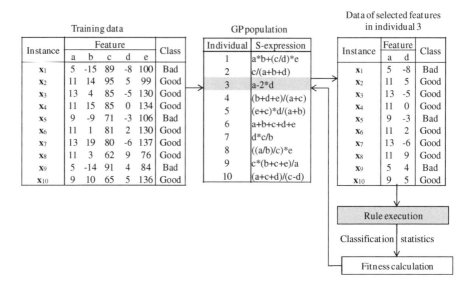

FIGURE 4.9: Fitness calculation in a typical GP.

Besides rule-based S-expressions, decision-tree-based representations are used where a decision tree consists of multiple rules [Folino *et al.*06b]. There are advantages and disadvantages of using single-rule S-expressions over decision trees. S-expressions can capture the complex relationships among the features, whereas representations of such complex expressions using simple decision trees are sometimes difficult. However, the downside of a single-rule S-expression is that it is limited to binary classification, and we need to decompose an n-ary classification problem into n binary classification problems, whereas, a single decision tree can be used to classify an n-ary classification problem. In this chapter, S-expressions are used to evolve rules for classification of binary as well as multiclass data.

4.4.2 Evaluation of a Classification Rule

The success of an evolutionary computation method, such as a genetic algorithm or a genetic programming is dependent on the fitness function used to evaluate an individual in the population. In a classification problem, by executing a classifier or a genetic programming rule on the training data, we get classification statistics of number of true positives (N_{TP}), true negatives (N_{TN}), false positives (N_{FP}), and false negatives (N_{FN}). In most classification problems, these classification statistics are combined in various ways to get a single score of goodness of an individual. In these sections, various fitness evaluation methods are described in detail. The steps in calculation of fitness of an individual in GA and GP are shown in Figs. 4.6 and 4.9.

Before we describe various fitness functions, let us define sensitivity, specificity, precision, and recall that will be used later. In the context of binary classification, sensitivity, specificity, recall, and precision of a target class are defined as follows:

$$sensitivity = recall = \frac{N_{TP}}{(N_{TP} + N_{FN})}; \quad (4.21)$$

$$specificity = \frac{N_{TN}}{(N_{TN} + N_{FP})}; \quad (4.22)$$

$$precision = \frac{N_{TP}}{(N_{TP} + N_{FP})}. \quad (4.23)$$

Sensitivity and specificity are used to measure how well a binary classification test correctly identifies the positive and negative cases, respectively. Precision is indicated by the portion of the cases with positive outcomes that are really positive cases. For example, in a medical test that determines if a person has prostate cancer, the sensitivity of the test to prostate cancer is the probability that the outcome of the test is positive if the person really has prostate cancer, whereas specificity is the probability that the outcome of the test is negative if the person does not have prostate cancer. Precision indicates the probability that in the case of a positive test outcome, the person really has prostate cancer. In medical screening tests, high sensitivity or specificity is required depending on the disease under test. For a contagious or infectious disease, high sensitivity is desired because early diagnosis and treatment may help prevent the spreading of the disease. However, if the wrong diagnosis of a disease has a negative effect on the patient psychologically and/or physically, high specificity is required. In safety management, high sensitivity is required because early detection of abnormality of a device or system may save the lives of human beings.

4.4.2.1 Accuracy-Based Fitness Function (A-fitness)

The straightforward method for calculation of fitness of a GP rule is the use of training accuracy. The accuracy of a rule is defined as the number of correct predictions (true positives + true negatives) divided by the total number of training instances. That is,

$$\text{A-fitness} = \frac{N_{TP} + N_{TN}}{N_{TP} + N_{TN} + N_{FP} + N_{FN}}. \quad (4.24)$$

Here, the fitness varies from 0 to 1. The main drawback of accuracy-based fitness function for classification of data is that the fitness is influenced by the class that has a higher number of instances in the training data.

4.4.2.2 Correlation-Based Fitness Function (C-fitness)

Matthews [Matthews75] proposed correlation between the prediction and the observed reality as the measure of raw fitness of a predicting program.

For a binary classification problem, the correlation (C) is defined as follows:

$$C = \frac{N_{TP}N_{TN} - N_{FP}N_{FN}}{\sqrt{(N_{TN} + N_{FN})(N_{TN} + N_{FP})(N_{TP} + N_{FN})(N_{TP} + N_{FP})}}. \qquad (4.25)$$

When the denominator of equation (4.25) is 0, C is set to 0. The standardized fitness of a rule is calculated as follows:

$$\text{C-fitness} = \frac{1 + C}{2}. \qquad (4.26)$$

Since C ranges between –1 and +1, the standardized fitness ranges between 0 and +1, the higher values being the better and 1 being the best.

4.4.2.3 Wilcoxon-Statistics-Based Fitness Function (W-fitness)

A receiver operating characteristic (ROC) curve is a graphical plot of the sensitivity vs. (1–specificity) for a binary classifier system as its discrimination threshold is varied. The ROC curve can also be represented equivalently by plotting the fraction of true positives vs. the fraction of false positives. It shows the trade-off between sensitivity and specificity of a classifier. It has been recently introduced in machine learning and data mining to measure the goodness of a classifier.

Usually, the area under ROC curve (AUC), which is drawn by varying the discrimination threshold of a classifier, is used as a measure of performance of a classifier, and AUC is calculated by using the trapezoidal rule. For discrete classifiers that obtain various sensitivity and specificity values in different independent runs, AUC is not measured by employing the trapezoidal rule because outliers may produce misleading AUC. Hanley and Mcneil [Hanley and Mcneil82] have shown that nonparametric Wilcoxon statistics can be used to measure the area under ROC curve of a discrete classifier. In a binary classification problem, the AUC calculated by using the Wilcoxon statistics for a point in ROC space becomes as follows:

$$\text{AUC} = \frac{1}{2}(sensitivity + specificity). \qquad (4.27)$$

We use the term *W-fitness* to refer to the AUC for a discrete point in the ROC space. W-fitness ranges between 0 (when no instance is correctly classified) and 1 (when all instances are correctly classified). This fitness function balances the sensitivity and specificity equally. Note here that the fitness is the arithmetic mean of sensitivity and specificity.

4.4.2.4 Geometric-Mean-Based Fitness (G-fitness)

The geometric-mean-based fitness of a rule is defined as follows:

$$\text{G-fitness} = \sqrt{(sensitivity \times specificity)}. \qquad (4.28)$$

Like the W-fitness function, this fitness function gives equal weights to sensitivity and specificity.

4.4.2.5   Biased Fitness (B-fitness)

We define another fitness function called biased fitness that incorporates sensitivity and specificity information, but assigns higher weight to sensitivity or specificity that corresponds to the majority class. This fitness function is defined as follows:

$$\text{B-fitness} = \frac{1}{2} sensitivity \times specificity + Bias \qquad (4.29)$$

where

$$Bias = \begin{cases} sensitivity/2 \text{ If TargetClass is the majority class;} \\ specificity/2 \text{ If TargetClass is the minority class.} \end{cases}$$

The first term balances sensitivity and specificity equally, whereas the second term biases the fitness toward the majority class.

4.4.2.6   F-Measure Fitness (F-fitness)

The weighted harmonic mean of precision and recall is called the F-measure or balanced F-score. It is defined as follows:

$$\text{F-measure} = \frac{2 \times recall \times precision}{recall + precision}. \qquad (4.30)$$

This is also known as the F1 measure, because recall and precision are evenly weighted. Unlike the previous fitness functions, the score of F-fitness is dependent on the target class. On heavily unbalanced data, the score of F-fitness taking the majority class as the target class will be higher than that taking the minority class as the target class.

4.4.2.7   AUC Balanced (AUCB)

Among the previously described functions, W-fitness and G-fitness treat the sensitivity and specificity equally but do not take into account how much the two measures are balanced. Here, another function called AUC balanced (AUCB) [Paul *et al.*08a] is introduced. It is a function of AUC and the amount of balancing of sensitivity and specificity, which is defined as follows:

$$\begin{aligned} AUCB &= \text{AUC} \times \text{Balancing\_Factor} \\ &= \frac{1}{2}(sensitivity + specificity) \times \\ &\quad (1 - |sensitivity - specificity|). \end{aligned} \qquad (4.31)$$

This function ranges between 0 and 1. When either sensitivity or specificity is 0, AUCB will be 0; when both sensitivity and specificity are 1, AUCB will be 1.

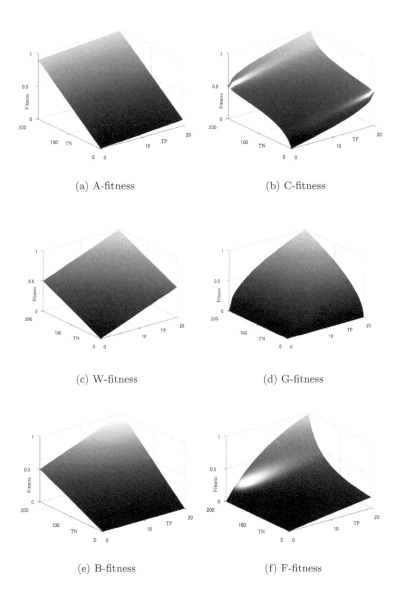

(a) A-fitness

(b) C-fitness

(c) W-fitness

(d) G-fitness

(e) B-fitness

(f) F-fitness

FIGURE 4.10: Various fitness functions.

4.4.2.8   Weighted Fitness

Though the classification statistics are widely used to evaluate the goodness of an individual, sometimes the number of selected features in an individual is used as a parameter of the fitness function. For example, the authors in [Liu and Iba01, Paul and Iba04b, Paul and Iba04a] use the weighted average of accuracy and the number of selected features as the fitness evaluation measure:

$$fitness(X) = w_1 \times \text{A-fitness} + w_2 \times (1 - d(X)/n) \qquad (4.32)$$

where w_1 and w_2 are weights from $[0, 1]$, $d(X)$ is the number of features selected in X, and n is the total number of features in the data set.

The fitness landscapes of the previously described six fitness functions (A-fitness, C-fitness, W-fitness, G-fitness, B-fitness, and F-fitness) are shown in Fig. 4.10. There are some observations to be made regarding the figures. For the C-fitness, W-fitness and G-fitness, the areas on both sides of the diagonal vertical plane passing through (0, 0, 0) and the optimum fitness (20, 200, 1) (or equivalently (1, 1, 1) when sensitivity and specificity are used) are the same. This means that the effects of the sensitivity and specificity on the fitness functions are the same. However, the score of G-fitness will be zero when either sensitivity or specificity is zero. When specificity is 1.0 and sensitivity is 0, the score of A-fitness will be 0.95, whereas those of B-fitness and G-fitness will be 0.5 and 0. Therefore, for classification of unbalanced data, B-fitness is better than A-fitness, and G-fitness is better than B-fitness. In a very unbalanced data set, F-fitness is the most difficult one to optimize; it is especially difficult to get a fitness value greater than 0.5 as can be seen from the figure in which the area above the horizontal plane at 0.5 is very small. For this fitness function, sensitivity must be greater than 0 to get a score better than 0.

4.4.3   Evolution of Multiple Rules for Multiclass Classification

By a single rule, we can classify data into two classes. For multiclass data, we need to evolve multiple rules for classification. There are two widely used approaches for classification of multiclass data: one-vs.-one and one-vs.-rest approaches.

In the one-vs.-one approach (also called an all-pairs approach), if there are c classes in the data, $c(c-1)/2$ rules are developed. Each $rule_{ij}$ is evolved using the training data from two classes (i, j), and the class of a test instance is predicted by the "winner-takes-all" voting strategy. If the rule says that the test instance is in class i, the vote for the i-th class is increased by one; otherwise the vote for j-th class is increased by one. Then the test instance is predicted to be in the class that has the highest votes.

In the one-vs.-rest approach, only c rules are evolved by c GP runs—one rule for each class. During evolution of a rule i, the instances of class i

are treated as positive instances, and other instances are treated as negative instances. In this case, the measures: true positive (TP), true negative (TN), false positive (FP), and false negative (FN) for calculation of the fitness of rule i are determined as follows:

$$\text{IF } (O(Y) \geq 0) \text{ AND } (\text{CLASS}(Y) = i) \text{ THEN TP;}$$
$$\text{IF } (O(Y) < 0) \text{ AND } (\text{CLASS}(Y) \neq i) \text{ THEN TN;}$$
$$\text{IF } (O(Y) \geq 0) \text{ AND } (\text{CLASS}(Y) \neq i) \text{ THEN FP;}$$
$$\text{IF } (O(Y) < 0) \text{ AND } (\text{CLASS}(Y) = i) \text{ THEN FN;}$$

where $O(Y)$ is the output of the S-expression of rule i on a test instance Y. Afterward, the rules are applied to the test instances to get generalized accuracy on them. If a GP is trained well, only one rule will fit an instance. If more than one rule fits an instance, the class is predicted either randomly or depending on the outputs of the rules. If the outputs of the rules are real values, the class of an instance is predicted to be the class that has the highest value of its rule. If two or more rules have the same value or outputs are Boolean values, the instance gets the label of the class that has the highest number of instances in the training data. The intuition behind this is that the class having the maximum number of training instances should always get priority. If none of the rules fit an instance, the instance is treated as misclassified (an error). An example of the one-vs.-rest approach for evolution of genetic programming rules for classification of multiclass instances is given in Fig. 4.11.

4.4.4   Limitations of Traditional GP Approach

The potential challenge for genetic programming is that it has to search two large spaces of functions and features simultaneously to find an optimal solution. Therefore, the proper choice of function set, genetic operators, and the depth of a rule is very important in applying GP to classification of data because increasing the depth of a rule and the number of functions may increase the complexity of the rules with little or no improvement in classification accuracy.

It has been observed that when the number of available training instances is very small compared to the number of features and the number of instances per class is not evenly distributed, a single rule or a single set of rules produces poor test accuracy. Because of a small number of training instances, most machine-learning methods use the LOOCV [Kohavi95] to calculate the training accuracy. Note that the data corresponding to the selected features remain the same during learning of the classifier using the LOOCV technique. Conversely, in genetic programming, different rules may evolve in different iterations of the LOOCV technique, and therefore we cannot calculate the

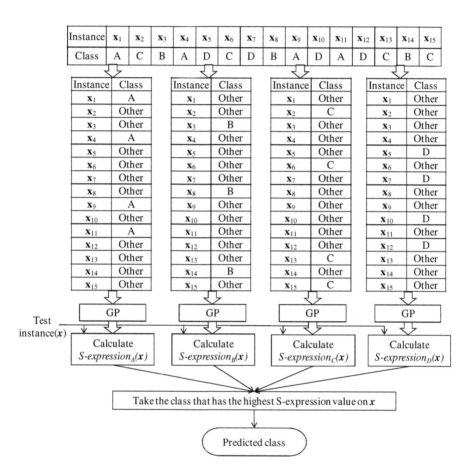

FIGURE 4.11: An example of the one-vs.-rest approach for evolution of genetic programming rules for classification of multiclass instances.

LOOCV accuracy of a particular rule. Instead, in most cases, the training accuracy of a rule is calculated by executing it on the whole training data in one pass. However, the single rules or sets of rules evolved in this way produce poor test accuracy.

4.5 Various Ensemble Techniques

In many data classification problems, single classifiers suffer from over-fitting. To improve the performance of single classifiers, various ensemble techniques have been proposed. These ensemble techniques use various methods to construct the training data, to build the classifier, and to predict the label of an instance with the ensemble. In this section, four ensemble techniques, namely, bagging [Breiman96], random forest [Breiman01], boosting [Schapire99], and MVGPC [Paul and Iba08, Paul *et al.*06] are described in detail.

4.5.1 Bagging

Bootstrap aggregating (bagging) [Breiman96] is an ensemble technique used to improve the stability and the accuracy of the base classifier. Given a training data set D of size N, a number of training subsets of size $N' \leq N$ are created by randomly sampling instances with replacement from D. Owing to sampling with replacement, some instances in each training subset may be duplicated. For a large data set, the size of the training subset is smaller than that of the original data but in a small data set, the training subset is the same size as the original data. Each training subset is used to build a classifier of the same type. Individual classifiers in the ensemble are combined by taking majority vote (for classification) or the average (for regression) of their outputs. An example of predicting the label of a test instance (**x**) using the bagging technique is given in Fig. 4.12.

4.5.2 Random Forest

The random forest classifier [Breiman01] utilizes the bagging and the random subspace [Ho98] to construct the individual decision trees in the ensemble. The class label of an instance is predicted using majority voting. Given a set of N training instances of n features, random forest first builds a number (= ensemble size) of training subsets by randomly sampling N instances with replacement from the training data. Then, each training subset is used to build a decision tree. During building of a decision tree, a number of features ($n' << n$) are randomly picked for each node of the tree and the best split

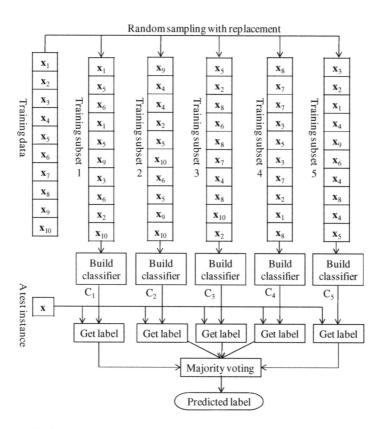

FIGURE 4.12: An example of bagging for classification.

is calculated using the training subset and the selected features. Each tree is fully grown and not pruned. The value of n' is held constant during the forest growing. An example of building a decision tree in the random forest classifier is given in Fig. 4.13. The main advantage of random forest is that it can build an ensemble very fast for a large database containing a large number of features.

4.5.3   Boosting

Boosting is a general method for improving the accuracy of any given learning algorithm [Schapire99]. The most widely known boosting algorithm is AdaBoost [Freund and Schapire97]. AdaBoost calls a given weak or base classifier repeatedly in a series of rounds and maintains a distribution or set of weights over the training set. Initially, all weights are set equally, but on each round, the weights of incorrectly classified examples are increased so that the weak learner is forced to focus on the hard examples in the training set. The final classifier is a weighted majority vote of the weak classifiers. The AdaBoost algorithm is given below.

Let $\mathbf{x}_i = (x_{i1}, x_{i2}, \ldots, x_{in})$ be the instance i where x_{ik} is the value of feature k in it and the label of this instance be $y_i \in \{0, 1\}$. The training set S is a collection of N labelled instances, i.e., $S = \{(\mathbf{x}_1, y_1), (\mathbf{x}_2, y_2), \ldots, (\mathbf{x}_N, y_N)\}$. Let $p_t^{(i)}$ represent the probability that the training instance i is included in the training set TR_t at iteration t; note here that $TR_t \subseteq S$. Let C be a genetic programming classifier that returns a rule $h_t : \mathbf{x} \rightarrow \{0, 1\}$ using the instances TR_t, and T be the maximum number of allowable iterations in AdaBoost. The steps in AdaBoost are as follows:

1. Set $t = 1$.
   Assign equal probability to each instance, i.e., $p_t^{(i)} = \frac{1}{N}$ for $i = 1, 2, \ldots, N$.

2. Pick N training instances with replacement using the probability distribution $\mathbf{p}_t$ to form TR_t.

3. Apply C to the instances $TR(t)$ to get a rule $h_t : X \rightarrow \{0, 1\}$.

4. Calculate error of h_t: $\varepsilon_t = \sum_{i=1}^{N} p_t^{(i)} |y_i - h_t(\mathbf{x}_i)|$.

5. If $\varepsilon_t > 0.5$, backtrack to previous iteration (set $t = t - 1$).

6. Calculate confidence level: $\alpha_t = \frac{1}{2} \ln \frac{1 - \varepsilon_t}{\varepsilon_t}$.

7. Update probability distribution:

$$
p_{t+1}^{(i)} = \begin{cases} \dfrac{p_t^{(i)} \exp(-\alpha_t)}{Z_t} & \text{if } y_i = h_t(\mathbf{x}_i); \\[2mm] \dfrac{p_t^{(i)} \exp(\alpha_t)}{Z_t} & \text{if } y_i \neq h_t(\mathbf{x}_i) \end{cases}
$$

where Z_t is a normalization factor so that $\sum_{i=1}^{N} p_{t+1}^{(i)} = 1$.

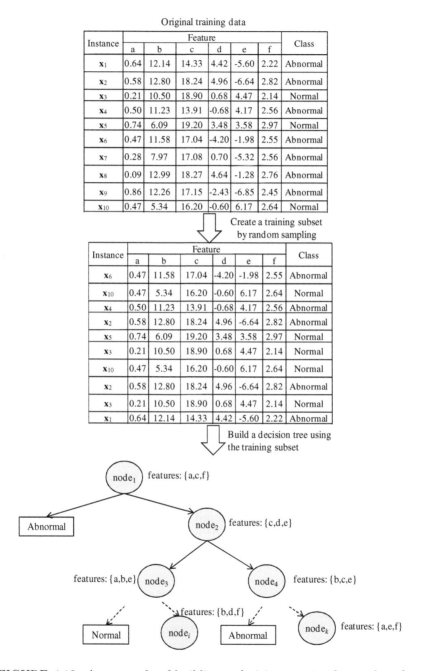

FIGURE 4.13: An example of building a decision tree in the random forest classifier.

8. $t = t + 1$. If $t < T$, go to **Step** 2.

9. Predict the label of an instance $\mathbf{x}$ as follows:

$$h_f(\mathbf{x}) = \begin{cases} 1 \text{ if } \alpha_+ > \alpha_-; \\ 0 \text{ otherwise} \end{cases}$$

where $\alpha_+ = \sum_{t=1}^{T} \alpha_t h_t(\mathbf{x})$ and $\alpha_- = \sum_{t=1}^{T} \alpha_t(1 - h_t(\mathbf{x}))$.

The basic difference between bagging and boosting is that in bagging, each instance has the same weight in the training data and each classifier has the same voting weight in the ensemble but in boosting, different instances have different weights in various iterations of the algorithm, and different classifiers have different weights in decision-making in the ensemble.

Though AdaBoost is widely used to improve the performance of a weak base classifier, it is sensitive to noisy data and outliers. Sometimes, it is no better than the base classifier. Moreover, for larger data sets, it sometimes fails to get a classification rule that has $\varepsilon_t > 0.5$.

4.5.4 Majority Voting Genetic Programming Classifier

Majority voting genetic programming classifier (MVGPC) [Paul and Iba08, Paul *et al.*06] is a simplified version of bagging and boosting. In this method, all the training instances are treated equally and rules are evolved using all the training instances in each GP run. That is, in each GP run, we are trying to evolve a rule that will perfectly classify all the training instances whereas in each run of AdaBoost, we are trying to evolve a rule that will perfectly classify a subset of training instances. Since all rules of the majority voting are evolved using all the training instances, their votes in prediction of the test labels are equally weighted ($\alpha_t = 1$). Fig. 4.14 shows the flow of steps in the MVGPC.

In MVGPC, the members of the ensemble are evolved independently using the whole training data. For binary classification, if v is the ensemble size, v genetic programming rules are produced in v independent runs. To predict the class label of a test instance, v GP rules are executed on the data of the test instance, the number of votes in favor of each class is counted and the class label is predicted to be the class that has the higher number of votes. To avoid a tie, v should be an odd number.

For multiclass classification, if v is the ensemble size and c is the number of classes in the data, $v \times c$ genetic programming rules are produced in $v \times c$ independent runs—v rules for each class. During evolution of a rule for a class $C_i(i = 1, 2, \ldots, c)$, one-vs.-rest approach is applied, in which all the instances from class C_i are treated as positive instances, and the remaining instances as negative instances. That is, each rule is a binary classifier. To predict the class label of a test instance Y, the rules in the ensemble are executed one by one on the test instance and the vote in favor (positive vote) and against

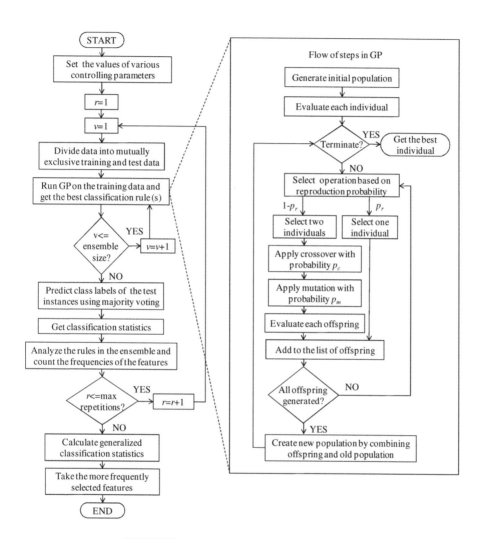

FIGURE 4.14: Flow of steps in MVGPC.

(negative vote) a class is counted. Then, the class label of the test instance Y is predicted as follows:

$$Class(Y) = \arg\max_{C_i}\{r_{C_1}, r_{C_2}, \ldots, r_{C_c}\} \qquad (4.33)$$

where r_{C_i} is the ratio of positive and negative votes of class C_i. If the number of negative votes of class C_i is zero, r_{C_i} is set to v. The test instance gets the label of the class that has the highest ratio of positive and negative votes. If two or more ratios are the same, the class is determined by randomly picking one class from the classes that have the same ratios. If all ratios are zero, the test instance is treated as misclassified. Let us give an example. Assume that there are five classes (A, B, C, D, E) of instances in a data set, and the ensemble size is 7. If the number of positive and negative votes for the classes are $\{0, 4, 5, 5, 3\}$ and $\{7, 3, 2, 2, 4\}$, respectively, the predicted class label of the test instance will be either C or D (should be randomly chosen).

4.5.4.1 Dependency of MVGPC on the Performance of the Rules in an Ensemble

The performance of an ensemble is dependent on the performance of the individual classifiers. An ensemble will perform better than its individual rules, if the rules classify the training data with good accuracies and make errors on different instances in the training data. When all the rules in the ensemble are poor, the ensemble too performs poorly. Here, the probability that the ensemble of the MVGPC is no better than the individual rules is given for binary classification.

Suppose that v (should be an odd number) is the ensemble size, m is the test size, and p is the probability that a single rule in the ensemble makes a false prediction. For simplicity, assume that the false prediction rate of each rule in the ensemble is the same. Since the majority voting technique is used to make a decision about the class label of a test instance, the test instance is misclassified if more than $\lfloor\frac{v}{2}\rfloor$ rules make false predictions. The probability that k rules out of v make false prediction is given by $_vC_k p^k (1 - p)^{v-k}$. Therefore, the probability of a test instance being misclassified by the ensemble is

$$e = \sum_{k=\lceil\frac{v}{2}\rceil}^{v} {}_vC_k p^k (1 - p)^{v-k} . \qquad (4.34)$$

If $p < 0.5$, the probability of misclassification by the ensemble will be less than p, i.e., majority voting will be better than a single rule.

For m test instances, the expected number of false predictions by a single rule will be $\lceil mp \rceil$. So, the probability that the ensemble will make false predictions equal to $\lceil mp \rceil$ or more, i.e., the probability that the ensemble is

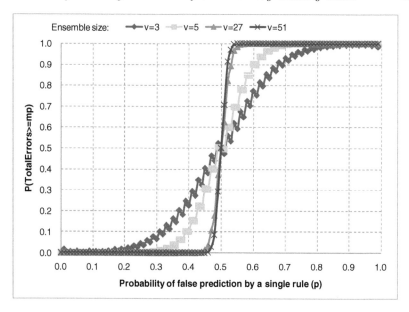

FIGURE 4.15: Probability that MVGPC will be worse than a single rule under $m = 51$ and different ensemble sizes (v).

not better than a single rule is

$$P(\text{Total Errors} \geq \lceil mp \rceil) = \sum_{j=\lceil mp \rceil}^{m} {}_{m}C_{j}\, e^{j}(1-e)^{m-j}. \qquad (4.35)$$

For a particular v and m, there is a cut-off value for p up to which MVGPC will be better than single rules, i.e., $P(\text{Total Errors} \geq \lceil mp \rceil)$ will be zero. As we increase v, the ensemble performs better for a larger cut-off value of p but that value is less than 0.5. Therefore, when $p < 0.5$, the test accuracy of MVGPC with a sufficient number of diverse rules in the voting group will be better than that of a single rule.

In Fig. 4.15, we have shown the graphical plots of the probability under $m = 51$ and different ensemble sizes (v). As we increase v, the cut-off value for p increases. Interestingly, the performance of MVGPC with ensemble size 27 and 51 is almost identical.

4.5.4.2 Generation of the Rules in an Ensemble of MVGPC Using LOOCV

We have already said that we cannot calculate the LOOCV accuracy of a particular rule. However, the members of an ensemble in MVGPC can be evolved using LOOCV in the following way:

- Generate N (=number of training instances) rules in N GP runs.

- In each run $i(i = 1, 2, \ldots, N)$, leave instance i for validation and use the remaining $(N - 1)$ instances as training data. If the evolved best rule can correctly classify the left out one, add this best rule to the ensemble.

- Apply majority voting on the test data using the members of the ensemble.

Note that the number of members in an ensemble may be smaller than N.

4.5.4.3 Identification of Important Features Using MVGPC

For identification of important features, first a classifier is devised, which will obtain higher classification statistics, and then the evolved rules are analyzed to determine the most frequently occurring features, i.e., first classification, then feature selection. To get a more stable frequency distribution of selected features, MVGPC should be repeated on the data several times. It has been observed that some features are always selected whatever feature selection algorithms and classifiers are used. In classification of microarray data, these more frequently selected features may be either closely related with the target cancer class or junk features that are highly correlated with distinction of different training and test instances but have no biological significance.

4.5.5 Some Related Works on GP Ensembles

Since an ensemble of GP rules produces better classification accuracy than the individual rules, various GP ensembles have been used to solve various problems [Imamura *et al.*03, Zhang and Bhattacharyya04, Hong and Cho06, Abraham and Grosan06, Iba99, Folino *et al.*06a]. These ensemble techniques have used different methods for evolution of multiple members of an ensemble and different ways of combining the members of the ensemble to predict the class of a test instance.

In [Iba99], the whole GP population is divided into a set of subpopulations, each of which is evolved independently using bagging and boosting techniques. Then, an ensemble of individuals is built by taking the best individual from each subpopulation, and the output of the ensemble is determined by taking the weighted median of the outputs of the individuals in that ensemble. In [Imamura *et al.*03], N-version genetic programming (NVGP), a parallel genetic programming of the island model, is applied to improve the performance of fault-tolerant systems. NVGP creates an ensemble of p individuals by randomly picking an individual from each of the p independent islands of population. The fitness of the ensemble is measured using a majority voting technique. In [Zhang and Bhattacharyya04], a collection of relatively small training subsets of equal size is built by randomly picking instances with replacements from the very big training data set. Then one GP model is trained on each subset, and the best classifier of each GP model is combined through majority voting to classify new instances. In [Hong and Cho06], ensemble of genetic programming rules has been used to classify DNA microarray data. Each ensemble has been built using diverse genetic programming rules, and the diversity among different classification rules is measured by comparing the structure of those rules. The class of an instance is predicted using the fusion method. In [Folino *et al.*06a], cellular genetic programming for data classification (CGPC), a parallel genetic programming of a cellular model, is used to induce an ensemble of decision trees. Each cell of CGPC maintains a subpopulation of decision trees and a subset of training data to evaluate the decision trees; to create the subsets of training data, bagging and boosting techniques are applied. The ensemble is applied to classify new instances by using a simple majority voting algorithm. In [Grosan and Abraham06], an ensemble of multi-expression programming (MEP) and linear genetic programming (LGP) is formulated using a multiobjective evolutionary algorithm for prediction of stock market.

4.6   Applying MVGPC to Real-world Problems

4.6.1   Classification of Microarray Data

Our first application domain is biological data; specially, microarray gene expression data, which is sometimes referred to as "wide data." The objective is to develop a mathematical model that can predict the cancer class of a patient accurately and identify the potential biomarker genes. The main characteristic of this data is that the number of available training instances is very small compared to the number of features. In this section, the term *gene* is used in place of *feature* so that the text is consistent with the application domain.

For classification of microarray gene expression data, clustering, support vector machine, k-nearest neighbor classifier, the naive Bayes classifier, to name a few prominent techniques, have been used [Li *et al.*06b, Pan *et al.*04, Shen and Tan06, Nutt *et al.*03, Singh *et al.*02, Guyon *et al.*02, Alon *et al.*99, Alizadeh *et al.*00, Slonim *et al.*00, Ben-Dor *et al.*99, Ding00, Golub *et al.*99, Ramaswamy *et al.*01, Bhattacharjee *et al.*01]. Recently, genetic programming-based methods have been applied to the problem as the evolved rules provide insights into the quantitative or logical relationships among the selected genes [Zhang and Wong08, Hong and Cho06, Paul and Iba07, Paul *et al.*06, Paul and Iba06a, Mitra *et al.*06, Langdon and Buxton04, Driscoll *et al.*03].

In this section, the results of various GP- and non-GP-based methods for classification of microarray data are presented. The descriptions of the microarray data sets and the results (comparative test accuracies and the potential biomarker genes) are taken from [Paul and Iba08].

4.6.1.1   Preprocessing of MicroarrayData

Microarray data files generated by Affymetrix's GeneChip (see `http://www.affymetrix.com` for details about GeneChip) contain expression values of various genes under various conditions or in subjects, such as in cancer patients and healthy persons. The data in a microarray data file are organized in such a way that each column contains the expression levels of various genes in a sample (an instance), and each row contains the expression levels of a single gene in various samples (instances). In addition to the gene expression values, there are various tags such as P, M, or A associated with each value in the file, which are usually discarded. The negative gene expression values and the outliers are replaced from the file with suitable lower and upper thresholds. Then, those genes are selected that show significant variations across various samples. Afterward, the values are scaled by taking logarithm, applying a linear transformation or normalization technique; sometimes, a combination of two transformation techniques, such as logarithm and linear transformation, is used to scale the data.

4.6.1.2   Microarray Data Sets

To evaluate the performance of various classifiers on microarray data, four publicly available microarray data sets are used. These data sets include brain cancer [Nutt *et al.*03], prostate cancer [Singh *et al.*02], breast cancer [Hedenfalk *et al.*01], and lung carcinoma [Bhattacharjee *et al.*01] data sets. For each data set, if the training and test data are in two files, they are merged to get a single file, and if it is required, the data are then preprocessed. The preprocessed data of breast cancer and lung carcinoma are available from the links given below. The other two data sets have been preprocessed following the suggestions of the authors of these data sets. In each experiment, the processed data set was randomly divided into two mutually exclusive training and test subsets. For larger data sets like the prostate cancer and the lung carcinoma data sets, the ratio of training to test size was approximately 1:1. For smaller data sets like the brain cancer and the breast cancer data sets, the ratio of training to test size was approximately 2:1 as suggested in [Dudoit *et al.*02]. However, during random split of a data set, precautions were taken so that the desired ratio was maintained for each class of instances in the training and test subsets. If precautions are not taken, instances of some classes may be absent in either training or test subset, which, in turn, may cause larger over-fitting.

4.6.1.2.1   Brain cancer data   The brain cancer data set is a binary classification problem consisting of expression levels of 12,625 genes of 28 glioblastomas (GBMs) and 22 anaplastic oligodendrogliomas (AOs) divided into two sets of classic and non-classic gliomas. The complete sets of data were downloaded from `http://www-genome.wi.mit.edu/cancer/pub/glioma`, the clas-

sic and non-classic data sets were merged and preprocessed. After preprocessing, only 4,434 genes were left. Then the training and test subsets were
constructed by randomly dividing the 50 instances into approximately 2:1 ratio. The training subset consisted of 19 ($=\lceil 28 * 2/3 \rceil$) glioblastomas and 15
($=\lceil 22 * 2/3 \rceil$) anaplastic oligodendrogliomas instances, and the test subset
consisted of the remaining 16 instances.

4.6.1.2.2 Prostate cancer data The prostate cancer data set is also a
binary classification problem like the brain cancer data set but larger than
the brain cancer data set. It contains 52 prostate tumor (PT) and 50 non-
tumor prostate (normal)(NL) instances, and 12,600 genes. Raw data of the
initial set are available at `http://www-genome.wi.mit.edu/MPR/prostate`.
After preprocessing of the raw data, only 5,966 genes were left. The training
and test subsets were constructed by randomly dividing the instances into 1:1
ratio, and each contained 51 instances.

4.6.1.2.3 Breast cancer data The breast cancer data set is a three-
class classification problem containing gene expression levels of 3,226 genes
across 22 instances of BRAC1, BRAC2, and sporadic classes. The preprocessed breast cancer data set is available at `http://research.nhgri.nih.
gov/microarray/NEJM_Supplement/`. One instance in this data set labelled
"Sporadic/Meth.BRCA1" was treated as "BRAC1." Therefore, the numbers
of BRAC1, BRAC2, and sporadic instances in the data set were 8, 8, and
6, respectively. Since the number of available instances is small, the training
and test subsets were constructed by randomly dividing the instances into 2:1
ratio. The training subset contained 6 BRAC1, 6 BRAC2, and 4 sporadic
instances, and the test subset contained the remaining 6 instances.

4.6.1.2.4 Lung carcinoma data The lung carcinoma data set is a five-
class classification problem consisting of 139 lung adenocarcinomas (AD), 21
squamous (SQ) cell carcinoma cases, 20 pulmonary carcinoid (COID) tumors
and 6 small cell lung cancers (SCLC), as well as 17 normal lung (NL) instances. After preprocessing, only 3,312 genes were left out of 12,600. The
complete data sets are available at `http://research.dfci.harvard.edu/
meyersonlab/lungca.html`. The training and test subsets were constructed
by randomly dividing the 203 instances into 1:1 ratio. The training subset
contained 103 instances (AD:70, SQ:11, COID:10, SCLC:3, and NL:9), and
the test subset contained 100 instances (AD:69, SQ:10, COID:10, SCLC:3,
and NL:8).

4.6.1.3 Test Accuracies

Here the results of various classifiers on the microarray data sets taken from
[Paul and Iba08] are presented. The results are shown in Tables 4.3 and 4.4.
In Table 4.3, the comparative test accuracies obtained by applying different

methods on the four data sets are summarized. On each data set, 20 independent experiments were performed. The ensemble sizes in MVGPC and AdaBoost+GP on the brain cancer, prostate cancer, breast cancer, and lung carcinoma data sets were 17, 27, 16, and 21, respectively. The values of various genetic programming parameters were: *population size* = 4000; *offspring size* = 3999; *maximum number of nodes in a GP tree (maximum rule size)* = 100; *maximum number of generations in a run* = 50; *maximum crossover depth* = 7; *maximum initial depth* = 6; *crossover probability* = 0.9; *mutation probability* = 0.1; and *reproduction probability* = 0.1. As a fitness function, C-fitness was used. Perfect training rule (PTR) is defined as a classification rule of genetic programming that can perfectly classify all the training instances in the data set and thus has a fitness of 1.0. A perfect set of training rules (PSTR) is defined as a collection of perfect training rules—one for each class—for classification of multiclass data. The numbers of PTRs/PSTRs among the majority voting rules that produced the best average test accuracies on the brain cancer, prostate cancer, breast cancer, and lung carcinoma data were 192, 141, 320, and 31, respectively.

For RPMBGA and GA, the population size, offspring size, crossover and mutation probabilities (for GA), maximum number of generations, and maximum number of runs (= experiments in MVGPC) were the same as for MVGPC. For GA, uniform crossover was used to create offspring as the number of bits in an individual is huge. Though the initial population of both methods was randomly generated, the maximum number of genes selected in each individual of GA was restricted to 100 (= the maximum number of nodes in a tree of a rule of MVGPC) because without this restriction, the fitness of GA did not improve over initial generation. The fitness of a gene subset was evaluated using the same method as used in [Paul and Iba05]. For k-NN with RPMBGA, various experiments were performed on the four data sets using 3, 5, and 7 nearest neighbor members in the k-NN classifier. In terms of average training and test accuracies, the best results were obtained on the brain cancer, prostate cancer, breast cancer, and lung carcinoma data with k = 3, 3, 5, and 3, respectively. For SVM with GA, we used LIBSVM [Chang and Lin01] implementation of SVM with C = 32, and γ = 0.0078125.

From Tables 4.3 and 4.4, we find that on the data sets MVGPC obtains significantly higher (at 5% level of non-parametric t-test) test accuracies than other methods. This evidence strongly suggests that MVGPC is an appropriate method for the prediction of the class label of a test instance from microarray data.

4.6.1.4   Potential Biomarker Genes

The more frequently selected genes are determined by analyzing the rules of the ensembles in the 20 experiments of MVGPC. In Table 4.5, those more frequently selected genes are presented, which are taken from [Paul and Iba08]. In the table, the official symbol (if any) and the name of a gene are given

Table 4.3: Comparative test accuracies on the data sets [Paul and Iba08, Table 1].

Data set Method	Brain cancer	Prostate cancer	Breast cancer	Lung carcinoma
MVGPC	80.31±5.08	90.59±2.07	79.17±16.11	95.50±1.54
AdaBoost+GP	66.56±12.04	82.35±4.97	67.50±13.76	92.65±2.30
PTR/PSTR	67.90±10.86	79.21±6.67	32.24±18.23	75.55±5.66
k-NN+RPMBGA	67.50±10.26	84.41±4.99	60.0±15.67	89.35±3.34
SVM+GA	56.25±0.0	51.37±1.97	65.0±23.51	76.80±1.51

Table 4.4: p-Values in statistical tests of significance at 5% level of non-parametric t-test (MVGPC vs. other method) [Paul and Iba08, Table 2].

Data set Method	Brain cancer	Prostate cancer	Breast cancer	Lung carcinoma
AdaBoost+GP	6.49E-06	7.43E-09	9.30E-03	1.11E-04
PTR/PSTR	2.98E-12	2.21E-36	6.22E-36	6.54E-17
k-NN+RPMBGA	1.72E-06	1.01E-07	2.92E-04	6.31E-11
SVM+GA	7.71E-13	6.16E-15	2.06E-02	9.04E-14

in the format *symbol: name* under the column "Gene description." The roles of these more frequent genes are checked by using the NCBI Entrez Gene Database (http://www.ncbi.nlm.nih.gov/sites/entrez) and those genes that are known to be associated with the types of cancer considered in this article are described in the texts.

Out of 4434 genes, 2355 genes were included at least once in the 340 rules of brain cancer data. Of these, the gene IGFBP2 [GenBank:X16302] was included the most times, 33, and is known to be associated with glioma progression, partly by enhancing MMP-2 gene transcription and, in turn, tumor cell invasion [Wang *et al.*03]. Among the other more frequently selected genes that are presented in the table, ECE1 [GenBank:Z35307] has a role in limiting Abeta accumulation in the mouse brain [Eckman *et al.*03]. The relationships of other frequently occurring genes with brain cancer are as yet unknown.

In the case of prostate cancer data, 3096 genes out of 5966 were selected at least once in the 540 classification rules. Of the five more frequently selected genes, HPN [GenBank:X07732] is known to have roles in prostate cancer progression. HPN is functionally linked to hepatocyte growth factor/MET pathway, which may contribute to prostate cancer progression [Kirchhofer *et al.*05].

In the 960 rules of breast cancer data, 3038 genes out of 3226 were selected at least once. Some of these more frequently selected genes are shown in the table. The four genes CRYAB [GenBank:NM_001885], PLAU [Gen-

Table 4.5: More frequently selected genes of the data sets [Paul and Iba08, Table 3].

Data set	Accession#	Gene description	Freq.
	X16302	IGFBP2: insulin-like growth factor binding protein 2, 36kDa	33
	M80254	PPIF:peptidylprolyl isomerase F	26
Brain	X77956	ID1: inhibitor of DNA binding 1	25
cancer	NM_005715	UST: uronyl-2-sulfotransferase	25
	AF007139	Homo sapiens clone 23898 unknown mRNA, partial cds	23
	Z35307	ECE1: endothelin converting enzyme 1	14
	M30894	Human T-cell receptor Ti rearranged gamma-chain mRNA V-J-C region, complete cds	126
Prostate	M84526	CFD: complement factor D (adipsin)	107
cancer	X07732	HPN: hepsin (transmembrane protease, serine 1)	85
	AL049969	Homo sapiens mRNA; cDNA DKFZp564A072	51
	D83018	NELL2: NEL-like 2 (chicken)	49
	NM_005749	TOB1: transducer of ERBB2, 1	18
Breast	NM_001885	CRYAB: crystallin, alpha B	17
cancer	NM_002658	PLAU: plasminogen activator, urokinase	15
	NM_001826	CKS1B: CDC28 protein kinase regulatory subunit 1B	14
	NM_053056	CCND1: cyclin D1	13
	NM_000424	KRT5: keratin 5 (epidermolysis bullosa, simplex Dowling-Meara /Kobner/Weber-Cockayne types)	142
	NM_001306	CLDN3: claudin 3	124
Lung	NM_003722	TP73L: tumor protein p73-like	119
carcinoma	NM_001005862	ERBB2: neuroblastoma/glioblastoma derived oncogene homolog	97
	NM_001305	CLDN4: claudin 4	89
	NM_002202	ISL1: ISL1 transcription factor, LIM/homeodomain, (islet-1)	76
	NM_006907	PYCR1:pyrroline-5-carboxylate reductase 1	73

Bank:NM_002658], CKS1B [GenBank:NM_001826], and CCND1 [GenBank: NM_053056] are of biological interest because they are known to be associated with breast cancer.

Among the more frequently occurring genes of lung carcinoma data, TP73L [GenBank:NM_003722] and ERBB2 [GenBank:NM_001005862] are known to have some roles in lung cancer. TP73L is consistently expressed in the squamous cell carcinoma in the lung, but non-consistently expressed in a subset of adenocarcinomas and large cell carcinomas [Au *et al.*04]; overexpression of ERBB2 is associated with recurrent non-small cell lung cancer [Onn *et al.*04].

However, it has been observed that the best rules or the sets of the best rules, which individually produce the highest test accuracies on a microarray data set, contain very few of these more frequently selected genes. This phenomenon leads to two assumptions. The first assumption is that the irrelevant or redundant genes in the best rule negatively affect the acquired classification accuracies of these more frequently selected genes. The second assumption is that the genes in the best rule are correlated and jointly affect the development or the suppression of a cancer; however, as single genes they are not differentially expressed in various types of tumor instances. Further studies are needed in this regard to reach to a concrete conclusion.

4.6.1.5   Speed of Convergence of MVGPC on Microarray Data

The speed of convergence of an evolutionary computation method to the optimum fitness depends on a number of factors including the population size, the number of good individuals in the initial population, the number of instances in the training data, the distribution of the instances and the complexity of the instances. Due to randomness in genetic programming, the execution times required to produce two classification rules for the same data may differ. However, for a given setting of genetic programming, it is expected that the average number of generations required by GP to reach the optimum fitness will increase with the increasing training size. For example, the average numbers of generations required by GP to produce a perfect training rule on prostate cancer data were 25.45 and 29.90 for training sizes of 51 and 68, respectively.

On a multi-class data set, when rules are evolved using the one-vs.-rest approach, the speed of convergence may follow not only the training size but also the complexity of the positive instances relative to other instances in the training data in a run. Sometimes, it takes a higher number of generations to produce a rule for a class containing a smaller number of instances than that for a class containing higher number of instances. For example, on lung carcinoma data, GP took on average 30.87, 9.68, 3.35, 2.42, and 5.61 generations to produce a perfect training rule for AD, SQ, SCLC, COID, and NL, respectively. Though SCLC has the smallest number of instances in the training data, the average number of generations required by GP to produce a perfect training rule for COID class is smaller than that required

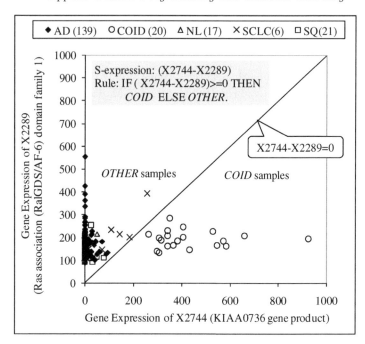

FIGURE 4.16: Plot of gene expression levels of X2744 and X2289 across various instances of lung carcinoma.

by GP to produce a perfect training rule for SCLC. This suggests that the COID instances are relatively easy to separate from instances of other classes whereas instances of other classes are not easily separable from one another [Paul and Iba08]. In Fig 4.16, the gene expression values of two typical genes X2744 (38032_at) (KIAA0736 gene product) and X2289 (39601_at) (Ras association (RalGDS/AF-6) domain family 1) across various instances of the lung carcinoma data set are plotted to show that COID instances are easily separable from other instances. The same observation had been made in many other classification rules of COID class.

4.6.2   Classification of Unbalanced Data

On microarray data, we have found that MVGPC outperforms other methods in terms of accuracy. In balanced data, each class has almost the same number of training instances, and simple accuracy is a good indicator of performance measurement of various classifiers. On the contrary, in unbalanced data, one class has a much larger number of instances than other classes, and simple accuracy is a useless index of performance measurement of a classifier, because in unbalanced data the accuracy is heavily biased by the majority class, which results in very unbalanced sensitivity and specificity.

Table 4.6: UCI machine-learning data sets used in the experiments.

Data set	#Features	#Majority instances	#Minority instances
WBC	30	357	212
MAGIC gamma	10	12332	6688
E. coli	7	301	35
Yeast	8	1240	244

To cope with the unbalanced data, various data resizing or weight adjusting techniques have been used in the past during the learning of a classifier [Li *et al.*06a, Sun *et al.*06, Zhang *et al.*04, Murphey *et al.*04, Domingos99, Ling and Li98, Kubat and Matwin97, Bruzzone and Serpico97]. Most of these methods reduce the size of the data by under sampling the majority class or increase the size of the data by over sampling the minority class [Ling and Li98, Kubat and Matwin97]. In either case, the misclassification rate of the minority class will decrease but the misclassification rate of the majority class will increase. When the size of the data is not changed, the weights or the misclassification costs of the majority and the minority classes are adjusted; usually higher cost is set to the misclassification of an instance from the minority class [Domingos99, Sun *et al.*06].

In this section, GP, MVGPC, AdaBoost+GP, and other non-GP classifiers are evaluated in term of AUC balanced (AUCB) [Paul *et al.*08a], which is a function of AUC and the balancing power of sensitivity and specificity. To evolve the rules of the ensembles of MVGPC and AdaBoost+GP, W-fitness and G-fitness are used.

4.6.2.1 Data Sets

To evaluate the performance of various classifiers on unbalanced data, six unbalanced data sets from different domains having distinct characteristics are chosen. Of these six data sets, four data sets are publicly available online and two data sets, which are collected from two Japanese companies, are not publicly available. These data sets are described below.

4.6.2.1.1 UCI machine-learning data sets UCI (University of California, Irvine) KDD (knowledge discovery in databases) archive data (URL: http://archive.ics.uci.edu/ml/) [Newman *et al.*98] are widely used to validate the performance of various machine-learning classifiers. The main characteristic of these databases is that the number of available training instances is usually much larger than the number of features. To validate the performance of various classifiers in this chapter, we have selected four data sets from the UCI KDD archive. These data sets are Wisconsin breast cancer (WBC) data, MAGIC gamma telescope data, E. coli data, and yeast data.

The descriptions of these data sets are given in Table 4.6. The E.coli and the yeast data are multi-class classification problems; we convert each of them into a binary classification problem by taking one class as the minority class, and the union of the remaining classes as the majority class. In the E. coli data set, the 35 instances of the imU (inner membrane, uncleavable signal sequence) class form the minority class, and the 301 instances of the remaining seven classes form the majority class. In the yeast data set, the 244 instances of MIT (mitochondrial) class form the minority class, and the 1240 instances of the remaining nine classes form the majority class. We have constructed training and test subsets by randomly splitting the instances of the minority and the majority classes into 1:1 ratio in each experiment. Therefore, the numbers of instances of the minority and the majority classes in the training subsets of WBC data, MAGIC gamma telescope data, E. coli data, and yeast data are (106, 179), (3344, 6166), (18, 151), and (122, 620), respectively.

4.6.2.1.2   Customer satisfaction data   This data set contains data on customer satisfaction (CS) about a product of a Japanese company. Customers were asked 84 questions in different categories about the product and they were asked to answer either agree (1) or disagree (0) depending on whether they agree with the statement about the product or not. Finally, they were asked about their overall evaluation (good or bad) of the product. Of those who responded, about 10% regarded the product as bad while about 90% regarded the product as good. In this article, we use the 84 questions as the features and the overall evaluations as the class labels to design a classifier to classify the data. We have constructed the training and test subsets by randomly splitting the whole data into 1:1 ratio in each experiment.

4.6.2.1.3   Financial data: credit risk management data   Commercial banks and many consumer finance companies offer a wide array of consumer financial services, ranging from issuing credit cards to providing small business loans. These banks and companies employ various techniques to promote repayment. Nonetheless, some of these financial institutions become bankrupt because of bad debts.

Though consumer finance companies usually have a lower volume of bad debts than traditional banks, the number of loan defaulters is not zero. Recently, to enhance profitability by lowering bad debts, some of these consumer finance companies have started applying data mining techniques to consumers' information. The objective of these techniques is to decide based on the personal data of an applicant whether he/she should be granted a loan.

The consumer data used in this paper have been collected from a leading consumer finance company in Japan that is interested in applying data mining techniques for credit risk management (CRM). This data set contains information of 32,422 consumers. Each individual is labelled as either normal or defaulter and has been summarized by 19 features. (Owing to privacy con-

Table 4.7: Settings of various parameters in the experiments.

Parameter	Value
Population size	1000 or 4000
Maximum generations	50
Total runs/repetitions	20
Reproduction rate	0.1
Crossover rate	0.9
Mutation rate	0.1
Maximum nodes in a GP tree	100
Maximum initial depth	6
Maximum crossover depth	7
Ensemble size (MVGPC)	11
#Iterations (AdaBoost)	11

cerns, neither the name of the finance company nor any of the features can be made public.) Among those consumers, only 3,003 (9.26%) are loan default-ers; therefore, these data are very unbalanced. For data mining purposes, this data set has been divided by the company into fixed training and test subsets. The training subset contains 14,614 normal and 1,392 defaulter instances, and the test subset contains 14,805 normal and 1,611 defaulter instances.

4.6.2.2   Results

The values of various genetic programming parameters for classification of unbalanced data are shown in Table 4.7. For smaller size data sets (WBC, E. coli, yeast, and CS data), the population size is set to 1000 while for larger size data sets (MAGIC gamma telescope data, and financial data), the population size is set to 4000. The maximum number of nodes in a GP tree is set to 100. As functions, we use logical functions (AND, OR, NOT, =, <>) during evolution of GP rules for classification of customer satisfaction data, and arithmetic functions (+, -, *, /) during evolution of GP rules for classification of the other five data sets. The division operator (/) is protected and $\frac{x}{0} = 1$. The algorithm terminates when either the best fitness of a generation is 1.0 or the maximum number of generations has passed. These same values of the parameters are used for evolution of the rules for MVGPC and AdaBoost+GP. For all data sets, the ensemble size of MVGPC and the number of iterations for AdaBoost+GP is set to an arbitrary value of 11. For all the data sets, the minority class is used as the target class and the classification statistics (sensitivity, specificity) are calculated accordingly.

The results of 20 repetitions are shown in Tables 4.8. In the table, a value of the form $a \pm s$ represents average value a with standard deviation s. To de-termine which of the three methods (basic GP, MVGPC, and AdaBoost+GP) performs the best using either of the fitness functions, statistical test of signif-

Table 4.8: AUCB obtained by MVGPC, GP, and AdaBoost+GP.

Data	W-fitness			G-fitness		
set	MVGPC	GP	AdaBoost	MVGPC	GP	AdaBoost
WBC	0.899	0.886	0.896	0.903	0.880	0.903
	±0.042	±0.043	±0.036	±0.036	±0.040	±0.037
MAGIC	0.743	0.709	0.637	0.746	0.724	0.688
	±0.011	±0.034	±0.073	±0.011	±0.024	±0.049
E. Coli	0.819	0.774	0.510	0.798	0.759	0.409
	±0.083	±0.105	±0.131	±0.067	±0.108	±0.135
Yeast	0.648	0.641	0.540	0.682	0.680	0.487
	±0.068	±0.072	±0.087	±0.047	±0.058	±0.079
CS	0.747	0.691	0.251	0.751	0.681	0.270
	±0.060	±0.085	±0.070	±0.046	±0.089	±0.087
Finan-	0.674	0.655	0.452	0.661	0.659	0.579
cial	±0.006	±0.018	±0.074	±0.009	±0.015	±0.029

icance (at 5% level of one-tailed Mann–Whitney test) on the average AUCBs of the three methods is performed.

The summary of the comparisons is presented in Table 4.9. In the table, the method having a significantly higher average AUCB than the other method is shown. In many cases, MVGPC obtains better results than GP and AdaBoost+GP using either of the fitness functions. On the contrary, AdaBoost+GP obtains worse results than GP in many cases. Using W-fitness and G-fitness, both MVGPC and GP obtain better AUCBs than AdaBoost+GP in 10 cases out of 12, but MVGPC obtains better AUCBs than GP in 7 cases out of 12. These results suggest that MVGPC is better than GP and AdaBoost+GP in classifying unbalanced data.

4.6.2.2.1 Performance of other non-GP classifiers To see how non-evolutionary classifiers perform on unbalanced data, support vector machine (SVM), C4.5 algorithm, and k-nearest neighbor classifier, which are widely used as benchmark classifiers, are applied to the data sets. To classify the unbalanced data sets, the Weka [Witten and Frank05] implementations of the algorithms are used. For SVM and C4.5, we use the default settings of the various parameters in Weka, and for the k-NN classifier, we perform experiments with $k=3$ and $k=5$ but $k=3$ produces better results in most of the experiments. On each data set, each classifier is run 20 times and evaluated using AUCB. The AUCBs obtained by these classifiers on the data sets are shown in Table 4.10. On the data sets, SVM produces poor AUCB than the other two classifiers in most of the cases. When these values are compared with the AUCBs obtained by GP and MVGPC using either W-fitness or G-fitness function (see Table 4.8), these classifiers are completely beaten by GP and

Table 4.9: Comparison of GP, MVGPC, and AdaBoost+GP in terms of average AUCB. The method returning significantly higher AUCB (at 5% level of one-tailed Mann–Whitney test) using a fitness function is shown in the table.

Data set	Methods	W-fitness	G-fitness
WBC	GP vs. MVGPC		MVGPC
	GP vs. AdaBoost		AdaBoost
	MVGPC vs. AdaBoost		
MAGIC	GP vs. MVGPC	MVGPC	MVGPC
	GP vs. AdaBoost	GP	GP
	MVGPC vs. AdaBoost	MVGPC	MVGPC
E. coli	GP vs. MVGPC	MVGPC	
	GP vs. AdaBoost	GP	GP
	MVGPC vs. AdaBoost	MVGPC	MVGPC
Yeast	GP vs. MVGPC		
	GP vs. AdaBoost	GP	GP
	MVGPC vs. AdaBoost	MVGPC	MVGPC
CS data	GP vs. MVGPC	MVGPC	MVGPC
	GP vs. AdaBoost	GP	GP
	MVGPC vs. AdaBoost	MVGPC	MVGPC
Financial	GP vs. MVGPC	MVGPC	
	GP vs. AdaBoost	GP	GP
	MVGPC vs. AdaBoost	MVGPC	MVGPC

MVGPC. Even in most of the cases, AdaBoost+GP produces better results than these three classifiers. The poor performance of these classifiers may be attributable to the way they learn the models from training data; since they treat each misclassified instance equally, the performance of the classifiers is biased by the majority class, which results in lower AUC as well as lower balancing factor. Therefore, balancing of the sensitivity and specificity during the learning of the classifier is very important. Genetic programming-based methods result in higher AUCB because they explore a huge number of candidate models and evaluate the models using a fitness function of sensitivity and specificity, which can be controlled.

4.6.2.2.2   Sensitivity and specificity of predictions by the classifiers
The average sensitivity and specificity of predictions by different classifiers using the six fitness functions are shown in Fig. 4.17. Among the six data sets, wide variations in sensitivity and specificity can be observed on all data sets except Wisconsin breast cancer data when A-fitness, C-fitness, B-fitness, or F-fitness is used to evolve the GP rules. As expected, basic GP and MVGPC balance sensitivity and specificity to some extent in the case of using either W-fitness or G-fitness on the five data sets (MAGIC gamma telescope data,

Table 4.10: AUCB by SVM, C4.5 and k-NN. A classifier returning significantly higher value (at 5% level of one-tailed Mann–Whitney test) than the other classifier is indicated by placing the initial letter of the other classifier as the superscript of the value of that classifier.

Data set	SVM	C4.5	k-NN
WBC	0.898±0.030	0.873±0.047	0.909±0.033^c
MAGIC	0.525±0.011	0.634±0.019^{s,k}	0.589±0.010^s
E. coli	0.0±0.0	0.394±0.186^s	0.508±0.129^s
Yeast	0.044±0.046	0.347±0.080^s	0.412±0.045^{s,c}
CS data	0.301±0.072^k	0.305±0.084^k	0.204±0.043
Financial	0.0±0.0	0.065±0.0^{s,k}	0.037±0.0^s

E. coli data, yeast data, customer satisfaction data, and financial data). One interesting observation is that the specificity of C-fitness is much higher than the sensitivity; that is, the score of C-fitness is influenced by the majority class like the A-fitness and the B-fitness. On Wisconsin breast cancer data sets, all the methods get very high sensitivity and specificity because the data set is easily classifiable and is relatively less unbalanced than the other five data sets. Interestingly, it appears that the fitness function has little impact on AdaBoost+GP because its performance using any of the six fitness functions does not change much.

4.7 Extension of MVGPC: Various Performance Improvement Techniques

The major challenge of applying MVGPC to a classification problem is the determination of the optimal ensemble size. When the ensemble size is very small, MVGPC does not produce better classification accuracies. With an increased ensemble size, MVGPC may produce better results but it will take very long execution time to produce all the rules in the ensemble. On a large, complex data set, GP may not reach the optimal fitness in every run of an experiment. In the worst case, the number of fitness evaluations per experiment will be $P \times G \times v \times c$ where P is the population size, G is the maximum number of generations per run, v is the number of rules per class, and c is the number of rules needed for classification of the data (for binary classification, $c = 1$), and in each fitness evaluation, a rule will be executed on all the training instances. For large-scale data mining problems, it may take a month or more to get the results of an experiment. In this section, various performance improvement techniques are discussed in detail.

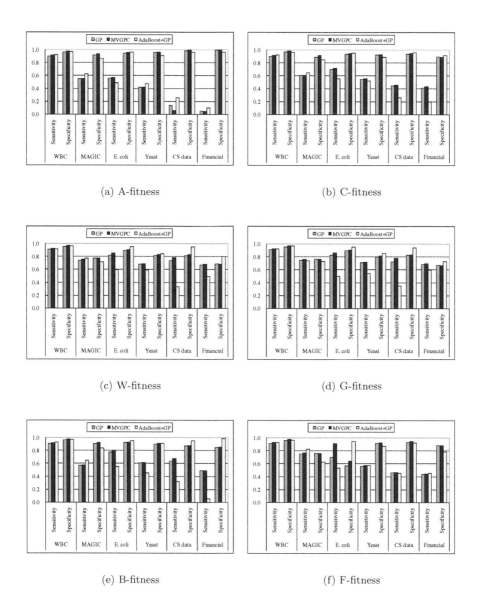

FIGURE 4.17: Sensitivity and specificity of classification of different methods on the data sets.

4.7.1   Class Prediction with Weighted Majority Voting Genetic Programming Classifier

In MVGPC, all the rules are equally weighted and the class of an instance is predicted using simple majority voting. However, because of randomness in genetic programming and a small number of training instances, in the case of microarray data, all the produced rules may not be equally strong on the training data. Some rules may be very strong, whereas some rules may be very weak. When the weak rules outnumber the strong rules, the performance of MVGPC will be no better than the weak rules (see Section 4.5.4.1 for details). To overcome this limitation, a framework as in Fig. 4.18 may be employed to predict the class of an instance. In this framework, during evolution of a GP rule, the training data are divided into internal training data and internal validation data; the internal training data are used to evolve the rules and the internal validation data are used to evaluate the best rule of a GP population. The classification statistics of the best rule on the internal validation data are used to calculate its weight in the ensemble of MVGPC. The higher the weight of a rule, the stronger the rule is. Afterward, the class of an instance is predicted using weighted majority voting. Note that this technique is different from both bagging and boosting. Unlike in bagging and boosting, neither the internal training data nor the internal validation data contain duplicate instances; that is, in weighted MVGPC:

Internal training data $\cup$ Internal validation data $=$ Training data; and

Internal training data $\cap$ Internal validation data $= \emptyset$.

Moreover, the method of calculation of weight of a rule is different from that in Adaboost.

4.7.2   Improvement of Accuracy by Heterogeneous Classifiers

In an ensemble containing heterogeneous classifiers, some of the members are produced with genetic programming, and some are produced with another non-evolutionary method, such as k-NN or SVM. The motivation behind it is that a non-evolutionary classifier, such as k-NN or SVM, takes shorter execution time to learn the classifier than GP and the reduction in the number of the GP rules in the ensemble will reduce the total execution time of an experiment and maintain diversity among the classifiers. During evolution of the GP rules, all the training instances are used but during the learning of a non-evolutionary classifier, a subset of instances selected randomly without replacement from the training data are used. Then, the class of an instance is predicted by using majority voting. An example of an ensemble of heterogeneous classifiers is shown in Fig. 4.19.

In Table 4.11, the classification accuracies obtained by applying heterogeneous classifiers and MVGPC are shown. During learning of an SVM classifier,

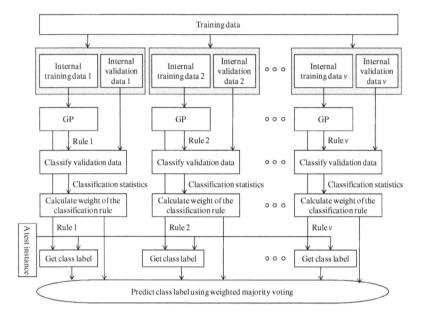

FIGURE 4.18: Class prediction with weighted majority voting genetic programming classifier.

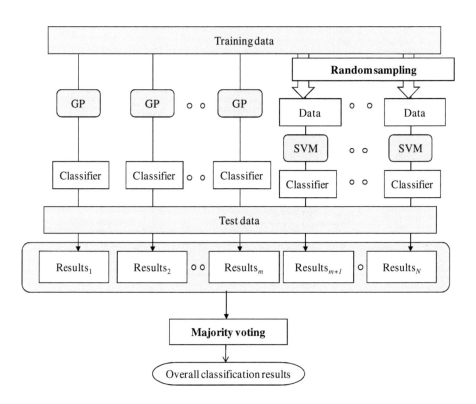

FIGURE 4.19: An ensemble of heterogeneous classifiers.

Table 4.11: Classification accuracies on the breast cancer data by the ensembles of heterogeneous classifiers and others.

Classifier	Average accuracy
GP	0.32
SVM	0.85
MVGPC	0.86
Ensemble of GP and SVM classifiers	0.87

13 instances are selected randomly from the training data of 17 instances (test data contain 5 instances). The ensemble is built with 15 GP rules and 15 SVM classifiers. For MVGPC, the ensemble size is set to 17. From the results, we see that the average accuracy is improved by 1%.

4.7.3 Improvement of Execution Time by Reducing Size of Training Data

Many real-world data contain huge numbers of instances and classification of those data by applying evolutionary computation methods is very costly and sometimes impractical unless the size of the training data is reduced. Moreover, some instances in a data set might be noisy and the performance of a classifier can be improved by removing those noisy instances from the data. Now the question is how to select an optimal set of representative instances from the data. A very small subset of instances will cause over-fitting. Various methods have been proposed for selection of instances (also called instance subset selection) from unbalanced data. Some of these methods are one-sided selections [Kubat and Matwin97], Wilson's editing [Barandela *et al.*04], synthetic minority oversampling technique (SMOTE) [Chawla *et al.*02], borderline SMOTE (BSM) [Han *et al.*05], and cluster-based over sampling [Jo and Japkowicz04]. Recently, the authors in [Hulse *et al.*07] have reported that the simple random under sampling (RUS) of the majority instances from the unbalanced data sets can produce reasonable classification results compared to other methods. Besides these algorithms, genetic algorithm can be applied to select a representative subset of instances from imbalanced data. To see whether GA obtains better results than RUS, GA is applied to four unbalanced data sets, namely, E. coli, yeast, customer satisfaction, and Wisconsin breast cancer data and the performance is measured using the cross-validation technique (see Fig. 4.20). As a fitness measure of GA, G-fitness is used and a classifier, k-NN with $k = 3$ is used. The results are shown in Table 4.12 and Figs. 4.21 and 4.22.

In Table 4.12, we find that RUS obtains significantly better (at 5% level of one-tailed Mann–Whitney test) AUCs on E. coli and yeast data but significantly lower AUCs on the CS data. On the WBC data, the AUCs obtained by GA and RUS are not significantly different from one other. In terms of

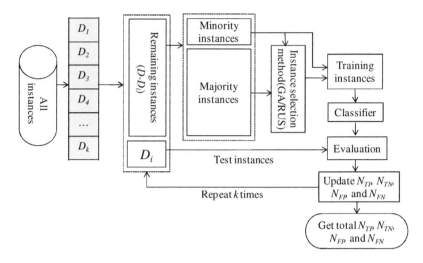

FIGURE 4.20: Evaluation of an instance subset selection method by using a cross-validation technique.

Table 4.12: Classification statistics of the instance subset selection by GA or RUS with k-NN classifier. All results are of 200 runs ($=$ 20 repetitions$\times$ 10-fold cross-validation). *Size* is the number of majority instances selected by a method; *Equal* means that the number of majority instances is equal to the number of minority instances while *Var.* indicates that the number of majority instances is $\pm 5\%$ that of the minority instances.

Data	Size	Method	Sensitivity	Specificity	G-mean	AUC
	Var.	GA	0.95±0.13	0.82±0.08	0.88±0.07	0.85±0.06
E. coli		RUS	0.91±0.17	0.86±0.07	0.88±0.10	0.88±0.09
	Equal	GA	0.95±0.12	0.82±0.07	0.88±0.07	0.85±0.06
		RUS	0.91±0.15	0.85±0.07	0.88±0.08	0.88±0.08
	Var.	GA	0.78±0.08	0.73±0.04	0.75±0.04	0.74±0.03
Yeast		RUS	0.76±0.10	0.75±0.04	0.75±0.05	0.75±0.05
	Equal	GA	0.78±0.07	0.73±0.04	0.76±0.04	0.74±0.03
		RUS	0.76±0.08	0.75±0.04	0.75±0.04	0.75±0.04
	Var.	GA	0.72±0.14	0.85±0.04	0.78±0.08	0.82±0.05
CS		RUS	0.67±0.15	0.87±0.04	0.76±0.09	0.77±0.08
	Equal	GA	0.71±0.15	0.86±0.04	0.78±0.08	0.82±0.05
		RUS	0.67±0.14	0.87±0.04	0.76±0.08	0.77±0.07
	Var.	GA	0.91±0.05	0.93±0.04	0.92±0.03	0.93±0.03
WBC		RUS	0.90±0.06	0.94±0.04	0.92±0.03	0.92±0.03
	Equal	GA	0.91±0.06	0.93±0.04	0.92±0.03	0.92±0.03
		RUS	0.90±0.07	0.94±0.04	0.92±0.04	0.92±0.04

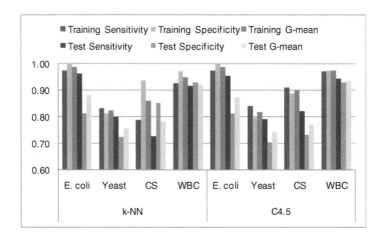

FIGURE 4.21: Performance of GA+k-NN on the training and test data.

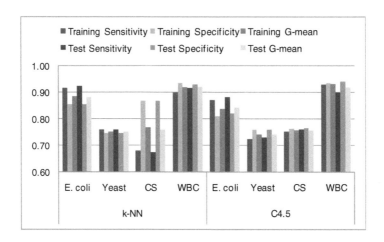

FIGURE 4.22: Performance of RUS+k-NN on the training and test data.

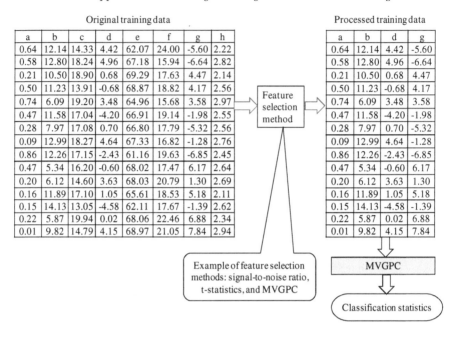

FIGURE 4.23: MVGPC on the data of the more frequently selected features.

G-means (G-fitness), RUS and GA are not significantly different from each other on all the data sets except on CS data where GA obtains better results than RUS. Moreover, in Figs. 4.21 and 4.22, we find that RUS suffers from less over-fitting than GA. These results imply that when the data set is very large, instead of a sophisticated learning algorithm, a simple random sampling may be employed to select a subset of representative instances from the data. This will reduce the time in the instance subset selection method as well as in the final learning method.

4.7.4 Reduction of Search Space of MVGPC

We also suggest that performance of the ensemble of rules containing only the more frequently selected features (see Fig. 4.23) be investigated because there is a possibility that this ensemble may produce better test accuracy. Here, MVGPC will be applied in two stages. In the first stage, MVGPC will be used as a preprocessor to identify more frequently occurring features. In the second stage, it will be applied to build a stronger predictive model using only the selected features in the previous stage.

Table 4.13: Notations used in calculation of diversity measure.

	C_j is correct	C_j is incorrect
C_i is correct	q	r
C_i is incorrect	s	u

4.7.5 Ensemble Building with Diverse GP Rules

To perform the ensemble better than the individual classifiers, it is required that the individual classifiers in the ensemble make errors on different instances. The intuition is that if each classifier makes errors on different instances, a strategic combination of these classifiers can reduce the total errors [Polikar06]. That is, we need the individual classifiers to be diverse from one another. Several measures have been used, such as correlation, Q-statistics, and entropy to calculate the diversity among the rules (see [Kuncheva and Whitaker01] for details).

Correlation diversity is measured as the correlation between the classification statistics of two classifiers, which is defined as

$$\rho_{i,j} = \frac{qu - rs}{\sqrt{(q+r)(q+s)(u+r)(u+s)}} \tag{4.36}$$

where q is the fraction of the instances that are correctly classified by both classifiers C_i and C_j, r is the fraction of the instances correctly classified by C_i but incorrectly classified by C_j, and so on (see Table 4.13). Note that $q+r+s+u = 1$. $\rho_{i,j}$ varies between –1 and +1, and when $\rho_{i,j} = 0$, maximum diversity is obtained.

Q-statistics [Yule1900] is defined as

$$Q_{ij} = \frac{qu - rs}{qu + rs}. \tag{4.37}$$

For statistically independent classifiers, $Q_{ij} = 0$. Q_{ij} varies between –1 and +1. Classifiers that tend to recognize the same instances correctly will have positive values of Q_{ij}, and those that commit errors on different objects will render Q_{ij} negative. For a set of v classifiers, the averaged Q_{ij} statistics of all pairs is taken as the measure of the diversity of the ensemble [Kuncheva and Whitaker01].

In the case of entropy measure, it is assumed that for an instance, the diversity is the highest among the v classifiers in the ensemble if half of the classifiers are correct and the remaining half are incorrect. If all the classifiers are either correct or incorrect, the classifiers are not diverse. Based on this assumption, the entropy is defined as

$$I = \frac{1}{N} \sum_{i=1}^{N} \frac{1}{v - \lceil v/2 \rceil} \min\{v, v - e_i\} \tag{4.38}$$

where N is the number of instances in the data, and e_i is the number of classifiers that misclassify the instance $\mathbf{x_i}$. I varies between 0 and 1; 0 means all the classifiers are the same and 1 means they are all diverse.

Given a diversity measure, various strategies can be employed to generate diverse classifiers for the ensemble. In one such strategy, one produces a sufficient number of classifiers, creates a number of ensembles by selecting randomly v classifiers, ranks those ensembles according to the diversity measure, and takes the top-ranked ensemble (see Fig. 4.24). In another strategy, one can use a combinatorial optimization technique, such as GA, to select an optimal subset of classifiers that produces better entropy. An example of such a strategy is shown in Fig. 4.25. In this strategy, the fitness of an ensemble is calculated using the diversity measure and the class label of a test instance is predicted using the weighted majority voting, where the weight of each rule in the best ensemble is calculated by using the classification statistics of the rule on the internal validation data. One can even employ a diversity measurement technique during the selection of the best rule from the GP population. In this case, instead of selecting the best rule from the GP population, one considers a number of rules with higher fitness and picks the one that will have the maximum diversity with the already selected classifiers in the ensemble. However, this method is dependent on the order of the rules that are picked from various GP populations.

4.8   Summary

In this chapter, various feature selection methods and classifiers, genetic programming, and various ensemble techniques were presented. The effectiveness of an ensemble technique called the majority voting genetic programming classifier (MVGPC) was demonstrated by applying it to various real-world problems, including microarray data, customer satisfaction, and credit risk management data. In most of the cases, it has been found that MVGPC is better than genetic programming, AdaBoost+GP, and other classifiers. Moreover, some of the more frequently selected genes in the ensemble of MVGPC are known to be potential biomarkers of cancer.

The success of evolutionary computation depends on various genetic parameters including the fitness function that is used to evaluate the goodness of an individual. In this chapter, various fitness evaluation techniques were discussed in detail. When the distribution of the classes in the data set is balanced, any one of the fitness functions can be used as an evaluation measure. However, for unbalanced data, it is necessary to take into account the sensitivity and specificity information during evaluation of fitness; some fitness functions, such as accuracy based fitness function (A-fitness) and correlation based

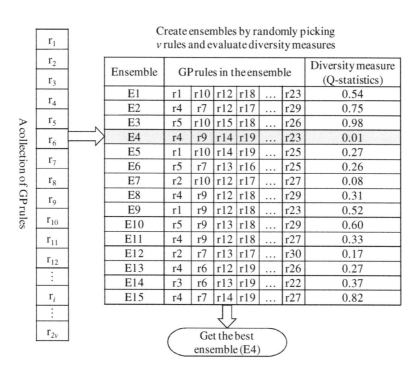

FIGURE 4.24: An example of building an ensemble with diverse rules.

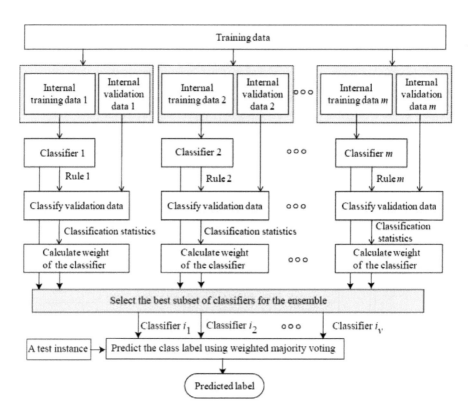

FIGURE 4.25: An example of building an ensemble with diverse rules and predicting the class label with weighted majority voting.

fitness function (C-fitness) are not good because the fitness is biased by the majority class. Instead of these fitness functions, Wilcoxon-statistics-based fitness, geometric-mean-based fitness functions, and AUC balanced (AUCB) should be used to measure the fitness.

The major limitation of applying evolutionary computation to feature selection and/or classification is that it requires very long execution time on larger data sets. If training data are downsized, the execution time will greatly decrease. In this chapter, it has been shown that instead of GA, simple random sampling can be deployed to select representative instances from the data set. Moreover, the execution time can be reduced by reducing the dimensionality of the data.

We have also shown that when the individual classifiers in an ensemble are very weak, the ensemble suffers from over-fitting resulting in poor test accuracy. Maintaining diversity among the individual classifiers is very important for creating a good ensemble; in this chapter, we have shown techniques to measure the diversity and create an ensemble with diverse GP rules. In MVGPC, all the individual classifiers are equally weighted, but the prediction strength of all rules may not be the same. A framework for prediction of the class label of an instance using weighted voting was also presented. In summary, a couple of improvement techniques presented in this chapter can be used to create a very strong ensemble for classification of large-scale balanced or unbalanced data.

Chapter 5

Probabilistic Program Evolution

5.1 Background

The optimization problems are to find the maximum or the minimum of functions given some constraints. In physics, this problem is dealt as an energy minimization problem. In optimization problems, many approaches have been proposed in several fields: mathematics, physics, and information sciences, for example. These approaches can been broadly classified into two approaches: deterministic and stochastic approaches. Stochastic optimization algorithms use random numbers. Among these methods, simulated annealing (SA) and evolutionary algorithm (EA) are popular. SA samples from the Gibbs distribution to find solutions. On the other hand, EA is based on the concept of natural evolution and applies genetic operators such as crossover and mutation to solution candidates. Although these algorithms are based on different concepts, EA can also be considered as an algorithm that samples from the distribution of promising solutions, with estimation of distribution algorithms (EDAs) attracting increasing attention [Larrañaga and Lozano01]. Because the distributions of promising solutions are unknown, an approximation is required to perform the sampling. Genetic algorithm (GA) [Holland75] and genetic programming (GP) [Koza92] sample from the distributions by randomly changing the promising solutions with genetic operators. On the other hand, EDAs assume a set of parametric models and sample from the predictive posterior distribution, according to

$$P(\mathbf{X}|\mathcal{D}) = \sum_{m\in\mathcal{M}} \int d\Theta P(\mathbf{X}|\Theta, m)P(\Theta|\mathcal{D}, m)P(m|\mathcal{D}), \qquad (5.1)$$

where $\mathcal{M}$ is a set of assumed models used in the EDA, Θ represents parameters, $\mathcal{D}$ represents learning data (the set of promising solutions), and $\mathbf{X}$ is an individual. EDA is similar to SA in the sense that both algorithms sample from parametric distributions. One of the most important features of EA is that EA not only improves from the viewpoint of the search performance, but also the applicability of the algorithm. EA has first been proposed as GA which uses fixed length linear arrays. However, it is very difficult to handle programs and functions using fixed length linear arrays. As a consequence,

GP, which adopts tree representations, has been proposed in order to make use of these structured data. Although problems which can be solved by GA are mainly concerned with parameter optimizations, GP can handle program and function optimizations. Since GP is an extension of GA, the traditional GP also uses genetic operators as crossover and mutation for evolving solution candidates. GA type mutation and crossover are analogies of the natural evolution. However, because GP chromosome is a tree structure which is quite different from that of the natural chromosome, GP is a less "naturally inspired algorithm". In recent years, EDA concepts have also been applied to GP type tree structures. In these new approaches, parametric models are assumed on tree structures and individuals are generated by sampling from estimated distributions. These algorithms are called GP-EDA or PMBGP (probabilistic model building GP). As mentioned previously, GP operators are less "genetic" and it is more natural to consider GP as an algorithm sampling from the distribution of unknown promising solutions as in EDA. In recent years, growing interests have been paid to the field of GP-EDA. EDA assumes a set of parametric models for distribution of promising solutions. In GA type linear chromosome, univariate model or Bayesian network is highly suitable for parametric models because GA only considers the fixed length chromosome and building blocks are position dependent. On the other hand, it is more difficult to assume good models in tree structures and this is a reason why GP-EDA has not been well-developed compared to GA type EDA.

Researches on GP-EDA can be broadly classified into two groups: prototype tree-based methods and probabilistic context-free grammar (PCFG)-based methods. The former methods translate variable length tree structures into fixed length structures (mostly linear arrays). The latter methods use the context-free grammar (CFG) for expressing programs and functions.

This chapter presents a congregate explanation on GP-EDA algorithms proposed up to now. We especially focus on POLE (program optimization with linkage estimation) and PAGE (programming with annotated grammar estimation), which are state-of-art GP-EDAs in prototype tree-based methods and PCFG-based methods, respectively. First, basics of EDA are explained in Section 5.1.1. We also give a brief explanation for several GP-EDAs proposed up to now, including both prototype tree and PCFG-based GP-EDAs.

5.1.1   Estimation of Distribution Algorithm

EAs are algorithms based on the concept of the natural evolution and, at the same time, are algorithms sampling from the promising distributions. Although crossover and mutation operators in GA and GP are mimics of those of living beings, it can be considered that these genetic operators are just approximation methods for sampling from unknown distributions.

Let $\mathcal{P}_g$ be a population at generation g and $\mathcal{D}_g$ be a set of promising solutions (throughout this chapter, $\mathcal{D}$ denotes data set). EA generally samples from distribution of $P(\mathbf{X}|\mathcal{D}_g)$, where $\mathbf{X}$ is an individual. However, in general,

---

Algorithm 1 General GA and GP.

---

1. $g \leftarrow 0$
   $\mathcal{P}_g \leftarrow$ Generate M random individuals

2. $\mathcal{D}_g \leftarrow$ Select $N \leq M$ promising individuals with the selection method

3. $\mathcal{P}_{g+1} \leftarrow$ Apply mutation and crossover to $\mathcal{D}_g$ to generate new population
   $g \leftarrow g + 1$

Algorithm 2 General EA.

---

1. $g \leftarrow 0$
   $\mathcal{P}_g \leftarrow$ Generate M random individuals

2. $\mathcal{D}_g \leftarrow$ Select $N \leq M$ promising individuals with the selection method

3. $\mathcal{P}_{g+1} \leftarrow$ Generate new individuals from a distribution of $\mathcal{D}_g$
   $g \leftarrow g + 1$

Algorithm 3 General EDA.

---

1. $g \leftarrow 0$
   $\mathcal{P}_g \leftarrow$ Generate M random individuals by $P(\mathbf{X}|\Theta_0, m_0)$

2. $\mathcal{D}_g \leftarrow$ Select $N \leq M$ promising individuals with the selection method

3. $\mathcal{P}_{g+1} \leftarrow$ Generate new individuals by sampling from a predictive posterior distribution

$$P(\mathbf{X}|\mathcal{D}_g) = \sum_{m \in \mathcal{M}} \int d\Theta \, P(\mathbf{X}|\Theta, m) P(\Theta, m|\mathcal{D}_g)$$

$g \leftarrow g + 1$

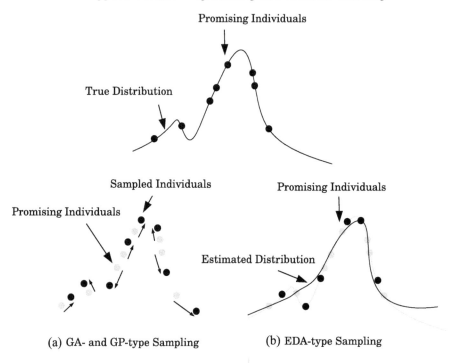

(a) GA- and GP-type Sampling   (b) EDA-type Sampling

FIGURE 5.1: GA- and GP-type sampling vs. EDA-type sampling.

the distribution $P(\mathbf{X}|\mathcal{D}_g)$ is unknown and an approximation is required to sample this distribution. GA and GP sample from the distribution by randomly changing the promising solutions. On the other hand, EDA assumes a set of parametric models and sample from the estimated distribution. Thus both GA (GP) and EDA sample from the distribution of promising solutions and only the difference between these two algorithms is an approximation method. In summary,

- GA and GP
  The sampling is performed by randomness (crossover and mutation).

- EDA
  A set of parametric models $\mathcal{M}$ is assumed and samples from the predictive posterior distribution given $\mathcal{D}_g$:

$$P(\mathbf{X}|\mathcal{D}_g) = \sum_{m \in \mathcal{M}} \int d\Theta\, P(\mathbf{X}|m, \Theta) P(\Theta|\mathcal{D}_g, m) P(m|\mathcal{D}_g).$$

Fig. 5.1 is an intuitive example which describes the difference between GA (GP) and EDA.

In the next section, we introduce basics of general EDAs.

5.2 General EDA

In this section, we explain the details of EDA. In Algorithms 3 and 1, we show the pseudocodes for general EDA and GA (GP), respectively. By comparing Algorithm 3 with Algorithm 2, which is a pseudocode for general EAs, we can see that EDA substitutes the sampling process. We explain the details of each operation of EDA in the following sections.

5.2.1 Initialization of Individuals

In the initialization phase, EDA generates random M individuals, where M is the population size. EDA generates new individuals from the distribution represented by equation (5.2):

$$P(\mathbf{X}|\Theta_0, m_0). \tag{5.2}$$

In this equation, Θ_0 is a set of initial parameters. For the case of binary GA-EDAs, each parameter is set uniformly. m_0 is an initial model and the simplest model is used in general. That is, if we consider Bayesian network for $\mathcal{M}$ (a set of parametric models), m_0 corresponds to an empty network structure (a network with no edges).

5.2.2 Selection of Promising Individuals

In this phase, promising individuals $\mathcal{D}_g$ are selected. These selected individuals are used for estimation of parameters and models (calculation of posterior of parameters and models). Any selection method can be used in EDA. A truncation selection is the most popular selection method in EDA. For the details of selection method, the readers may refer to Section 2.1, explaining basics of the conventional GP.

5.2.3 Parameter and Model Learning

EDA estimates the posterior probability $P(m|\mathcal{D}_g)$ using selected individuals $\mathcal{D}_g$ and calculates the predictive posterior distribution using the posterior (equation (5.3)),

$$P(m|\mathcal{D}_g) \propto P(m)P(\mathcal{D}_g|m), \tag{5.3}$$

where m is a parametric model. The posterior for parameters is also calculated in the same fashion when the model m is given (equation (5.4)):

$$P(\Theta|\mathcal{D}_g, m) \propto P(\Theta|m)P(\mathcal{D}_g|\Theta, m). \tag{5.4}$$

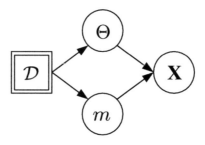

FIGURE 5.2: A graphical model representing predictive posterior distributions.

In these equations, $P(m)$ and $P(\Theta|m)$ are priors for m and Θ. Generally, uniform distribution ($P(m) \propto 1$, $P(\Theta|m) \propto 1$) is used. As mentioned later, the predictive posteriors are often approximated by MAP (maximum a posteriori) values due to computational reasons.

5.2.4   Generation of New Individuals

EAs generally generate new individuals from the distribution of promising individuals. EDAs assume a set of parametric models and approximate the distribution with these models. The predictive posterior of an individual $\mathbf{X}$ given data $\mathcal{D}_g$ can be represented by equation (5.5):

$$
\begin{aligned}
P(\mathbf{X}|\mathcal{D}_g) &= \sum_{m \in \mathcal{M}} \int \mathrm{d}\Theta \, P(\mathbf{X}|\Theta, m) P(\Theta, m|\mathcal{D}_g) \\
&= \sum_{m \in \mathcal{M}} \int \mathrm{d}\Theta \, P(\mathbf{X}|\Theta, m) P(\Theta|\mathcal{D}_g, m) P(m|\mathcal{D}_g).
\end{aligned} \tag{5.5}
$$

In this equation, $\mathcal{M}$ represents a set of parametric models. Fig. 5.2 describes a graphical model of the distribution. However, in reality, it is intractable to sum over possible models in equation (5.5). Furthermore, even if the number of models is small enough to be able to calculate summation, the predictive posterior distributions tend to be very complicated. As a consequence, we cannot use fast sampling methods such as PLS (probabilistic logic sampling), and most EDAs approximate the predictive posterior with MAP values (equation (5.6)),

$$
P(\mathbf{X}|\mathcal{D}_g) = P(\mathbf{X}|\widehat{\Theta}, \widehat{m}). \tag{5.6}
$$

In this equation, $\widehat{\Theta}$ is a MAP value and is represented by following equations:

$$
\widehat{m} = \operatorname*{argmax}_{m} P(m|\mathcal{D}_g),
$$

$$\widehat{\Theta} = \underset{\Theta}{\mathrm{argmax}}\, P(\Theta | \mathcal{D}_g, \widehat{m}).$$

The use of MAP values in the predictive posterior corresponds to approximating each posterior with

$$P(\Theta | \mathcal{D}_g, m) = \delta(\Theta - \widehat{\Theta})$$

$$P(m | \mathcal{D}_g) = \delta_{m,\widehat{m}} = \begin{cases} 1 & m = \widehat{m} \\ 0 & \text{otherwise} \end{cases}$$

in equation (5.5) ($\delta(x)$ is the delta function calculated $\int dx\, f(x)\delta(x-a) = f(a)$). When point values are used, a very fast sampling method such as PLS can be used in most of graphical models. If we used uniform priors over parameters $P(\Theta | m) \propto 1$, MAP values are identical to maximum likelihood estimators (MLEs). After generating new individuals, EDAs check the termination criteria. If the termination criteria are met, EDAs report the best individual. If not, EDA again performs the selection.

5.2.5 Estimation of Distribution vs. Genetic Operators

The advantages and drawbacks of these two methods (GA-type and EDA-type sampling) depend on the problems they are applied to. For the case of sampling based on randomness (GA and GP type sampling), two structurally similar individuals should have similar fitness values. The term "structural similarity" in this context may mean the Hamming distance, for example.

It is well known that GA can solve the OneMax problem very effectively, which is one of the most famous benchmark tests in GA. The fitness function of the OneMax problem is represented by equation (5.7):

$$fitness(\mathbf{X}) = \sum_{i=1}^{n} X_i. \tag{5.7}$$

In this equation, since GA uses a fixed length linear array, we assume that individual $\mathbf{X} = \bigcup_{i=1}^{n}\{X_i\}$ where X_i is a random variable at i-th position. In the OneMax problem, we can see that there is a linear relationship between difference of fitness and Hamming distance (structural similarity). As mentioned above, this problem can be solved effectively by GA. However, it is not expected that this assumption is always satisfied in general problems. The deceptive function of order k problem is a benchmark [Deb and Goldberg92] which violates the assumption. The fitness function of the deceptive function is represented by equation (5.8):

$$fitness(\mathbf{X}) = \sum_{j=1}^{s} f_{dec}^{k}(s_j), \tag{5.8}$$

Fitness landscape

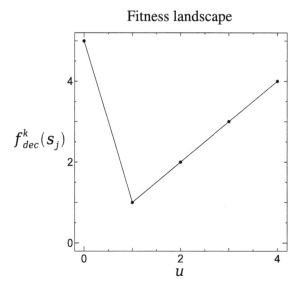

FIGURE 5.3: Fitness landscape of f_{dec}^k.

where s_j are non-overlapping subsets of $\mathbf{X}$ ($s_j \subset \mathbf{X}$, $s_i \cap s_j = \emptyset$ $(i \neq j)$) and s is the number of the subsets s_j. f_{dec}^k is a partial fitness function and is given by following values (these values are examples for the case of $k = 4$, where u is the number of 1s in 4 bits. Fig. 5.3 is a visualization of this function):

$$f_{dec}^4(s_j) = \begin{cases} 5 & u = 0 \\ 1 & u = 1 \\ 2 & u = 2 \\ 3 & u = 3 \\ 4 & u = 4 \end{cases}$$

The optimum of the deceptive problem can be obtained by combining 0000s. Although the Hamming distance between 0000 and 0001 is only 1, the fitness difference between these two strings is 4. This fact indicates that structurally similar individuals do not necessarily have similar fitness values. It is expected that it is difficult for GA to solve this problem. Actually, as the problem size becomes larger, it has been reported that GA requires much more fitness evaluations to obtain the optimum. On the other hand, one of the EDAs named BOA (Bayesian optimization algorithm) [Pelikan *et al.*99b] can solve this problem effectively. Because EDA does not take advantage of the assumption required in GA, this problem is not "deceptive" at all for BOA.

5.2.6 Several Benchmark Tests for GP-EDAs

When using tree representations, several elements affect the search performance. Thus, when comparing traditional GP with proposing GP-EDAs, we have to show the effectiveness of the proposing methods not only quantitatively but also qualitatively. Even if new proposing methods show 10% better performance against GP, just showing the experimental data is not enough for reasoning the superiority of the methods. Because GP and GP-EDA have many different points, the difference of search performance may not be derived from key features of the proposing GP-EDAs. We have proposed two GP-EDAs, POLE (see Section 5.3.4) and PAGE (see Section 5.4.6) in this chapter. For clarifying the search behavior of these two methods, we used the following GP benchmark tests for the comparison.

- MAX problem [Gathercole and Ross96]

- Royal tree problem [Punch98]

- DMAX problem [Hasegawa and Iba06]

These benchmark tests have the common features listed below.

- Not having introns

- GP-hard problems

Current GP-EDAs cannot cope with introns. Although introns are redundant structures from the viewpoint of fitness evaluations, they are taken into account in model and parameter inferences. As a consequence, existence of introns greatly affects the search performance of GP-EDAs. The benchmark tests listed previously are considered to be GP-hard problems. We do not have to apply GP-EDAs to GP-easy problems. It is desirable for GP-EDA to solve the problems where traditional GP cannot. In this section, we briefly explain the above three benchmark tests (these benchmarks are used in Section 5.3.4.8 and Section 5.4.6.8).

5.2.6.1 MAX Problem

The MAX problem [Gathercole and Ross96, Langdon and Poli97] was proposed so as to take advantage of defects of a crossover operator in GP. As a consequence, the MAX problem has been widely used as a comparative benchmark test in GP-EDAs [Yanai and Iba03, Shan *et al.*04]. The objective of this problem is to search for the function which yields the largest value under constraints. In this problem, function and terminal nodes are limited to the nodes represented below:

$$\mathfrak{F} = \{+, \times\}, \quad \mathfrak{T} = \{0.5\}, \tag{5.9}$$

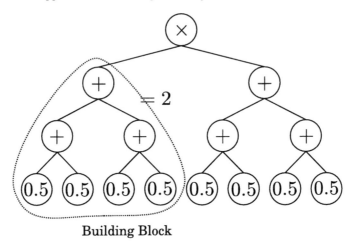

Building Block

FIGURE 5.4: The optimum of the MAX problem.

where $\mathfrak{T}$ and $\mathfrak{F}$ are a set of GP terminals and functions, respectively. Moreover, the MAX problem limits the maximum tree depth. An optimum of this problem can be obtained by creating building blocks which return "2" by using four "0.5"s and three "+" (see Fig. 5.4). By multiplying this 2 by using "×", the maximum value of the maximum depth D_P is given by $2^{2^{D_P-3}}$.

5.2.6.2 Royal Tree Problem

The royal tree problem [Punch98] is an extension of a well-known benchmark test named the royal road function [Mitchell *et al.*92] in GA. The royal road function was originally designed to investigate the schema hypothesis in GA. In a similar fashion, the royal tree problem is used for investigating the search behavior of GP. The optimum of the royal tree problem is combinations of smaller optima and is obtained by a crossover operator in GP. The royal tree problem employs a set of functions defined by alphabetical symbols $\mathfrak{F} = \{A, B, C, D, \cdots\}$, and the terminal nodes $\mathfrak{T} = \{x, y, \cdots\}$. The royal tree problem defines the *perfect tree* state. Fig. 5.5(a) describes examples of perfect trees.

A fitness value of an individual is given by the score of its root node ($fitness = Score(X_0)$), where $Score(X_i)$ is the score of X_i. The fitness calculation of the royal tree problem is carried recursively as,

$$Score(X_i) = wb_i \sum_j (wa_{ij} \times Score(X_{ij})), \qquad (5.10)$$

where X_{ij} is a j-th child (in a tree structure, counting from the left) of X_i, and wa_i and wb_{ij} are weights defined as follows:

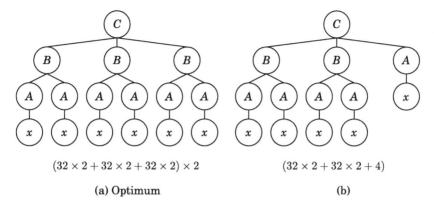

$$(32 \times 2 + 32 \times 2 + 32 \times 2) \times 2 \qquad\qquad (32 \times 2 + 32 \times 2 + 4)$$

(a) Optimum (b)

FIGURE 5.5: Examples of the scoring system for the royal tree problem. Equations in this figure represent the scoring calculations.

- wa_{ij}

  - *Full Bonus = 2*
    If a sub-tree rooted at X_{ij} has a correct root and is a perfect tree.
  - *Partial Bonus = 1*
    If a sub-tree rooted at X_{ij} has a correct root, but is not a perfect tree.
  - *Penalty = 1/3*
    If X_{ij} is not a correct root.

- wb_i

  - *Complete Bonus = 2*
    If a sub-tree rooted at X_i is a perfect tree.
  - *Otherwise = 1*

Fig. 5.5 shows examples of fitness calculations of the royal tree problem. The difficulty of this problem can be tuned by changing the number of function nodes. Generally, level E is used for comparisons since it is relatively difficult for GP to solve this level of the problem.

5.2.6.3 Deceptive MAX (DMAX) problem

The deceptive MAX problem (DMAX problem for short) is an extended problem of the MAX problem. Although the MAX problem takes advantage of the defect of crossover operator in GP, the MAX problem can be easily solved because the MAX problem does not have deceptive factors. In Section 5.2.4, we denoted that GP-type sampling is valid in the case that two structurally similar individuals have close fitness values. In the DMAX problem, this

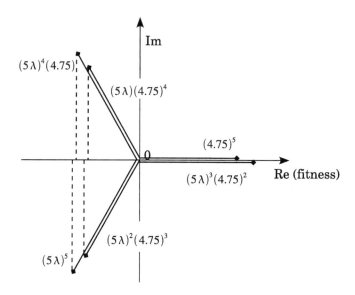

FIGURE 5.6: The optimum individual for the DMAX problem ($m = 5$, $r = 3$, $D_P = 3$) plotted in the complex plane.

assumption is not satisfied at all. As a result, it is very difficult for GP to obtain the optimum of the DMAX problem.

An objective of the DMAX problem is essentially identical to that of the MAX problem. The objective is to search for the functions which yields the largest real value under the maximum depth D_P using the given function and terminal nodes. The symbols used in the DMAX problem are different from those used in the MAX problem (equations (5.11), (5.12) and (5.13):

$$\mathfrak{F} = \{+_m, \times_m\},$$
$$\mathfrak{T} = \{\lambda, 0.95\}, \tag{5.11}$$

$$\lambda^r = 1, \lambda \in \mathbb{C}, r \in \mathbb{N}, \lambda \neq 1 \tag{5.12}$$

$$+_m(a_0, a_1, \cdots, a_{m-1}) \equiv \sum_{i=0}^{m-1} a_i,$$
$$\times_m(a_0, a_1, \cdots, a_{m-1}) \equiv \prod_{i=0}^{m-1} a_i. \tag{5.13}$$

In these equations, λ is a complex value satisfying the condition represented by equation (5.12). Since the DMAX problem employs a complex valued terminal, individuals in the DMAX problem generally yield complex values. Because the objective of the DMAX is to find the largest real value, the fitness value of the DMAX problem is given by equation (5.14):

$$fitness_i = \mathrm{Re}(individual_i). \tag{5.14}$$

Problem parameters are m (arity), r (power) and D_P (maximum tree depth). The difficulty of the DMAX problem can be tuned with these three parameters.

In this section, we explain the optimum value (the maximum value) for $m = 5$, $r = 3$, and $D_P = 3$ (this setting except for D_P is used in Sections 5.3.4.8 and 5.4.6.8). First, create 5λ by using five λs and $+_5$. Multiplying this structure by $\times_5$ yields $(5\lambda)^5 = 5^5\lambda^5 = 5^5\lambda^2$. Although the absolute value of $5^5\lambda^2$ is the largest, its real part is negative and is a negative solution. To make a real part of solutions the largest, two values out of the five values multiplied have to be real values. 5×0.95 are used instead for two 5λs, and this substitution gives the maximum value calculated by $(5\lambda)^3(0.95 \times 5)^2 = 2820.3125$. For the case of $D_P = 4$, the optimum is given by $(5\lambda)^{24}(0.95 \times 5) = 2.83 \times 10^{17}$.

Fig. 5.6 describes visualization in the complex plane of values which give large absolute values ($D_P = 3$). In this figure, horizontal axis represents the real axis, which is the fitness value at the same time. Therefore the value which is described at the right-most position is the optimum of the problem. Although the structures between $(5\lambda)^3(0.95 \times 5)^2$ (the optimum) and $(5\lambda)^4(0.95 \times 5)$ are very similar, the fitness values of these two individuals have quite different values. As can been seen, the assumption of GP-sampling is strongly violated. This fact indicates that the DMAX problem is highly inadequate for traditional GP.

5.2.7 Extending EDA to Tree Structures

In GA-EDA, it is relatively easy to apply parametric models since GA uses fixed length structures for chromosomes. From the viewpoint of statistical estimation, promising solutions $\mathcal{D}_g$ are learning data. Each chromosome can be represented by multivariate random variables. Because GA chromosomes are of fixed length, all the learning data have the same dimension. This property makes the task of model learning in GA-EDA very easy. At the same time, GA building blocks are position dependent. This feature enables the use of univariate model (mean-field type approximation) or the Bayesian network.

On the other hand, it is much more difficult to apply EDA techniques to GP because of the following reasons.

1. **Variable length chromosome**
   The statistical estimation in GA-EDA is relatively easy because GA adopts fixed length chromosomes. On the other hand, tree structures used in GP are variable lengths in general. This means that the dimension of each and every learning data is different and simply applying graphical models as Bayesian networks is impossible.

2. **Position independent building blocks**
   In tree representations, not only absolute positions of symbols, but also relative positions are very important. For example, if $\sin x$ is the optimum, $1 \times \sin x$ is also an optimum. These two optima have the same building block at different positions.

3. **Introns**
   Introns are structures that do not affect fitness values. However, these structures are taken into account in the probability distributions. It is very difficult to estimate the effect of introns without evaluating programs or functions. The most difficult problem we think is the effect of introns and as far as we know, no GP-EDAs solved this problem. This problem is discussed in Section 5.6.

Considering these problems, two types of GP-EDAs have been proposed.

- **Prototype tree-based method**
  This type of approach was first proposed in PIPE (probabilistic incremental program evolution). This method converts variable length tree structures into fixed length structures (e.g, fixed length linear arrays as GA), then GA-EDA approaches are applied to converted structures. This type of algorithm solves the first problem in the previous list.

- **Probabilistic context-free grammar (PCFG)-based method**
  GGGP (grammar guided GP) is the first method which employed the CFG (context-free grammar) for chromosomes. Stochastic grammar-based GP (SG-GP) extended GGGP to PCFG. In this method, all the functions and programs are expressed by derivation trees (see Section 5.4.1). Several PCFG-based algorithms solved the problem of the first (variable length chromosome) and the second (position independent building blocks) ones[1].

In the following sections, we first give the description of prototype tree-based GP-EDAs followed by those based on PCFG.

5.3 Prototype Tree-based Methods

When applying probabilistic models to tree structures, the simplest idea is to convert tree structures to linear fixed length arrays and apply probabilistic models to these arrays. Since chromosomes of GA-EDAs are fixed length, techniques devised in the field of GA-EDAs can be applied after the

---

[1] However, there are many PCFGs whose models are position dependent.

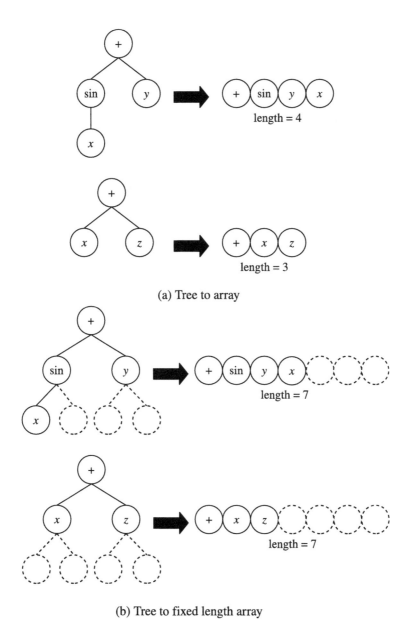

(a) Tree to array

(b) Tree to fixed length array

FIGURE 5.7: Conversion of tree structures to fixed length structures.

Table 5.1: Comparison of prototype tree-based GP-EDA. EPT denotes probabilistic prototype tree and expanded parse tree, respectively.

	Graphical Model	Chromosome	Estimation of Interaction	Adaptive Model Learning
PIPE	Univariate	Prototype tree	No	No
EDP	Parent-child	Prototype tree	Yes	No
XEDP	Vertical relation	Prototype tree	Yes	No
ECGP	Joint distribution	Prototype tree	Yes	Yes
POLE	Bayesian network	EPT	Yes	Yes

conversion. As mentioned in the previous section, if the length of each chromosome is the same, it is relatively easy to carry a statistical estimation. This type of approach was first introduced in PIPE and this approach is called the prototype tree-based approach. For example, Fig. 5.7(a) is an example of a conversion of tree structures to arrays using the breadth-first traversal. By using this conversion, any tree structures can be converted to linear arrays. However, because the length of the converted linear array depends on the size of original tree structures, the length differs among individuals. It is impossible to apply GA-EDA probabilistic models to variable length arrays. In order to overcome the difficulty, most prototype tree-based GP-EDAs consider full trees to make all the converted arrays the same length. In Fig. 5.7(b), dummy nodes are added to make the tree structures full trees. The advantage of prototype tree approaches is that it is easy to import the techniques devised in GA-EDA. Furthermore, prototype tree approaches are very intuitive and relatively easy to implement. However, it is quite difficult for prototype tree-based approaches to deal with position independent building blocks, which might be important in GP.

In the following sections, we briefly introduce several prototype tree-based GP-EDAs (Table 5.1). After these explanations, we will make a detailed explanation for our proposed method POLE (program optimization with linkage estimation). Throughout this section, we assume $\mathbf{X} = \bigcup_{i=1}^{n}\{X_i\}$, where X_i is a random variable of i-th node in prototype trees.

5.3.1   PIPE

The first method, which has introduced prototype trees in the field of GP-EDA, is algorithm named PIPE (probabilistic incremental program evolution) [Sałustowicz and Schmidhuber97a, Sałustowicz and Schmidhuber97b]. PIPE uses the probabilistic prototype tree (PPT) to handle programs and functions. PIPE assumes that there are no dependencies among nodes (equation (5.15)):

$$P(X_1, X_2, \cdots, X_n) = \prod_i P(X_i). \qquad (5.15)$$

In each node, probabilities of symbols (function or terminal) are stored (see Fig. 5.8). PIPE generates new individuals based on this decomposed

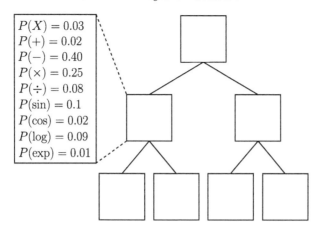

$$P(X) = 0.03$$
$$P(+) = 0.02$$
$$P(-) = 0.40$$
$$P(\times) = 0.25$$
$$P(\div) = 0.08$$
$$P(\sin) = 0.1$$
$$P(\cos) = 0.02$$
$$P(\log) = 0.09$$
$$P(\exp) = 0.01$$

FIGURE 5.8: PPT in PIPE. Probabilities of symbols are given in each node.

distribution. The parameter learning of PIPE is an incremental learning. The parameters are updated to make the probability that the best individual in the population generated is higher. Since, PBIL (population-based incremental learning) [Baluja94] uses the similar parameter learning method in GA-EDA, it is considered that PIPE is an extension of PBIL to GP. Because the model of PIPE is very simple, the cost of parameter learning is not expensive. On the other hand, because PIPE does not take into account the interactions between nodes, PIPE is not suitable for the problem where there are strong interactions between nodes.

5.3.2 ECGP

ECGP (extended compact GP) [Sastry and Goldberg03] is an extension of ECGA (extended compact GA) [Harik99], which is a GA-EDA using marginal distributions for grasping dependencies of the building blocks. By using marginal distributions, interactions between nodes can be taken into account.

ECGA assumes that the full joint distribution can be expressed by products of smaller marginal distributions (equation (5.16)):

$$P(X_1, X_2, \cdots, X_n | \mathcal{D}_g) = \prod_{s \in \Omega} P(s | \mathcal{D}_g). \tag{5.16}$$

In this equation, Ω is a set whose elements are $s \subset \mathbf{X} = \bigcup_{i=1}^{n} \{X_i\}$, where $s_1, s_2 \in \Omega(s_1 \neq s_2) \Rightarrow s_1 \cap s_2 = \emptyset$. In short, ECGA assumes that the joint distribution is decomposed into products of non-overlapping marginal distributions. If Ω is a set of schemata, $s \in \Omega$ represents schema. By expressing the joint distribution with partial marginal distributions, the schema can be preserved.

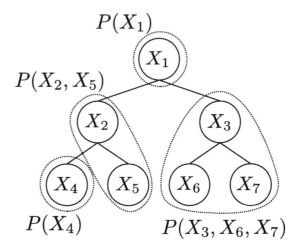

FIGURE 5.9: Probabilistic model of ECGP. Building blocks are represented with marginal distributions.

In ECGA, Ω has to be estimated using learning data (promising solutions). ECGA adopted MDL (minimum description length) [Rissanen78] principle to infer Ω. In particular, MDL of ECGA can be expressed by equation (5.17):

$$\text{MDL} = f(N) \sum_{i=1}^{m_{\text{prt}}} (r_i^{k_i} - 1) + N \sum_{i=1}^{m_{\text{prt}}} \sum_{j=1}^{r_i^{k_i}} (-p_{ij} \log p_{ij}). \qquad (5.17)$$

In equation (5.17), m_{prt} is the number of partitions, k_i is the number of variables in i-th partition, r_i is the number of cardinalities of X_i ($r_i = ||X_i||$), and N is the number of learning samples. The first part in equation (5.17) corresponds to the number of parameters and the second part to the likelihood of learning data. A function $f(N)$ depends on information criteria and $f(N) = \log N$ in MDL.

ECGP applies the concept of ECGA to prototype trees. Fig. 5.9 is an example of a target tree (programs or functions). If ECGP learns the set of schema represented by Fig. 5.9, then this means the whole distribution is decomposed as the following equation:

$$P(X_1, X_2, \cdots, X_7 | \mathcal{D}_g) = P(X_1 | \mathcal{D}_g) P(X_2, X_5 | \mathcal{D}_g) P(X_4 | \mathcal{D}_g) P(X_3, X_6, X_7 | \mathcal{D}_g).$$

Because ECGP preserves the information of interactions between nodes, ECGP is more suitable than PIPE in problems which exhibit strong dependencies among nodes. However, ECGP can only grasp relatively small building blocks because of the parameter size limitation. The parameter size of one marginal distribution can be represented by r^k, where r is the number

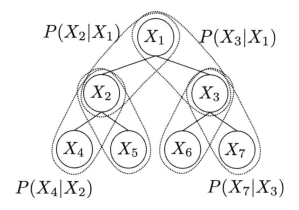

$P(X_2|X_1)$ $P(X_3|X_1)$

$P(X_4|X_2)$ $P(X_7|X_3)$

FIGURE 5.10: Probabilistic model of EDP.

of possible symbols and k is the size of the target marginal distribution. We can see that the size of parameters is exponential to k. For the case of GA, $r = 2$ (0 and 1 are only used). On the other hand, GP uses more symbols in general (e.g, $r = 10$). Thus if we consider the marginal distribution of size 3, this marginal distribution requires $\sim 10^3$ parameters ($r = 10$ case). Larger parameters require larger population size and as a result, more fitness evaluations are required.

5.3.3 EDP

GA-EDAs have to infer the graphical model with the maximum posterior probability (which expresses the underlying relationships) since GA chromosomes do not have explicit graph structures. On the other hand, GP chromosomes are tree structures and it is very natural to think that it may be effective to use tree structures themselves as expressing node interactions. In general, the graph structure learning is very computationally expensive task and if this assumption works well, a use of the tree structures for grasping the interactions among nodes may be highly computationally effective approach.

EDP [Yanai and Iba03] takes advantage of this idea. Relationships among nodes are fixed during search in EDP. The distribution of EDP can be expressed by equation (5.18):

$$P(X_1, X_2, \cdots, X_n) =$$
$$P(X_1) \prod_{X_{ch} \in ch(X_1)} P(X_{ch}|X_1) \prod_{X'_{ch} \in ch(X_{ch})} P(X'_{ch}|X_{ch}) \cdots . \quad (5.18)$$

In this equation, X_i is i-th node and $ch(X_j)$ is a set of children nodes of X_j in the tree structure (see Fig. 5.10).

A parameter estimation in EDP is basically MLE. However, to avoid the over-fitting problem and the probabilities of symbols with zero frequencies, EDP infers the parameters according to the equations below:

$$P' = \eta \widehat{P} + (1 - \eta)P_g,$$

$$P_{g+1} = (1 - \alpha)P' + \alpha \frac{1}{|\mathfrak{F}| + |\mathfrak{T}|}.$$

In these equations, $\widehat{P}$ is MLE, P_{g+1} is a probability at generation g, $|\mathfrak{F}| + |\mathfrak{T}|$ is the number of symbols (terminals and functions), η is the degree of a dependence on parameters of a previous generation, and α is a parameter for the Laplace correction. The effect of the Laplace correction can be taken as a mutation operation. Other EDAs also consider similar effect. For example, if Bayes estimation is adopted, prior distribution will avoid the over-fitting and symbols with probabilities 0.

XEDP (extended EDP) [Yanai and Iba05] is an extended method of EDP which tries to grasp the position-independent building blocks. XEDP additionally estimates the conditional probabilities described with Fig. 5.11. Intuitively, this can be represented by the equation below:

$$R(\text{Children}|\text{Parent}, \text{Grandparent}).$$

This probability distribution is called "recursive distribution" in XEDP. The parameters of recursive distribution are estimated in the same fashion:

$$R' = \eta \widehat{R} + (1 - \eta)R_g,$$

$$R_{g+1} = (1 - \alpha)R' + \alpha \frac{1}{|\mathfrak{F}| + |\mathfrak{T}|}.$$

Since XEDP estimates fixed building blocks, XEDP is not a computationally expensive method. XEDP is reported to be able to solve many GP benchmarks (e.g, the regression problem, the parity problem, and the wall following problem) with fewer fitness evaluations than GP.

5.3.4 POLE

In the previous section, we gave a brief explanation of several prototype tree-based GP-EDAs. We next show a detailed explanation for a state-of-art prototype tree-based GP-EDA named POLE (program optimization with linkage estimation) [Hasegawa and Iba08]. POLE employs a Bayesian network for graphical models. By using a special chromosome called the *expanded parse tree* (EPT), POLE overcomes the problem caused by the use of a probabilistic model for GP. The results of this section are taken from [Hasegawa and Iba08].

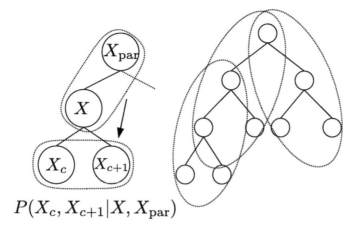

$$P(X_c, X_{c+1}|X, X_{\text{par}})$$

FIGURE 5.11: A probabilistic model of XEDP.

5.3.4.1 Introduction

In previous works of prototype tree-based GP-EDAs, univariate and adjacent models have been applied. However, in order to estimate more general interactions between nodes, Bayesian network is considered to be more suitable. However, there is a problem in applying Bayesian network to prototype tree-based GP-EDA. Bayesian network is widely used in GA-EDA because of its model flexibility. For the case of GA-type binary chromosomes, the number of possible symbols is 2 (0 and 1). On the other hand, for the case of GP, more nodes are required to express functions and programs. As a result, conditional probability table (CPT) which expresses the quantitative relationship among nodes becomes larger. This gives rise to increase of the number of learning samples. Furthermore, all the programs and functions generated (sampled) from Bayesian network must be syntactically correct.

In order to overcome the above problems, POLE employs a special chromosome called the expanded parse tree. The characteristics of our approach are listed below.

- A Bayesian network is used for estimating the interactions between nodes.

- A special chromosome called the expanded parse tree is employed to reduce the number of possible symbols.

We first introduce the basics of Bayesian network and we then make a detailed explanation of POLE.

5.3.4.2   Bayesian Network

In this section, we make a brief explanation for Bayesian network (see [Heckerman95] for details).

When large data sets are given, it is very important to model the data to extract information from them. Because data contain uncertainty in general, non-deterministic expressions using probabilities are preferred to deterministic ones such as decision trees. The Bayesian network is a graphical model using graphs and probabilities to express the relationship of variables. By using Bayesian network, we can describe the non-deterministic cause and effect among events intuitively. The Bayesian network is a highly popular graphical model because of its model expressiveness. Furthermore, Bayesian network is a directed acyclic graph (DAG) and sampling from the network is not computationally expensive. This advantage is useful for EDA because sampling process is a key feature in generating new individuals in EDA.

Suppose we are given a set of n variables $\mathbf{X} = \bigcup_{i=1}^{n}\{X_i\}$. Bayesian network expresses the joint probability of $\mathbf{X}$. By using the Bayes chain rule, the joint distribution can be decomposed as equation (5.19):

$$P(X_1, X_2, \cdots, X_n) = \prod_{i=1}^{n} P(X_i|X_1, \cdots, X_{i-1}). \qquad (5.19)$$

Describing the relationship in equation (5.19), we obtain Fig. 5.12. In this figure, every node is connected and we cannot say that this graphical model expresses the data intuitively. Thus consider a subset $\Pi_i \subset \mathbf{X}$, which satisfies the equation below:

$$P(X_i|X_1, X_2, \cdots X_n) = P(X_i|\Pi_i). \qquad (5.20)$$

This conditional independence decides a structure of Bayesian network. In Bayesian network, Π_i corresponds to a set of parent nodes of X_i. By using equation (5.20), the joint probability distribution can be expressed by equation (5.21):

$$P(X_1, X_2, \cdots, X_n) = \prod_{i=1}^{n} P(X_i|\Pi_i). \qquad (5.21)$$

For example in $\Pi_5 = \{1, 2, 3\}, \Pi_4 = \emptyset, \Pi_3 = \{1, 2\}, \Pi_2 = \emptyset$, Bayesian network can be described by Fig. 5.13.

5.3.4.3   Bayesian Network Learning

There are two types of learning in Bayesian network: one is parameter and the other is structure. Parameter learning (estimation) in Bayesian network is carried out in Bayesian fashion. That is to calculate the posterior distribution of parameters Θ given data $\mathcal{D}$ and structure G (equation (5.22)):

$$P(\Theta|\mathcal{D}, G) \propto P(\mathcal{D}|\Theta, G)P(\Theta|G). \qquad (5.22)$$

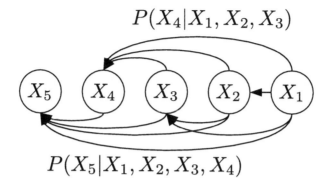

FIGURE 5.12: A Bayesian network of equation (5.19). Arrows $X_i \to X_j$ represent the conditional dependency $P(X_j|X_i)$.

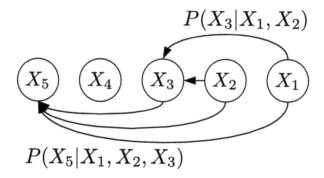

FIGURE 5.13: An example of Bayesian network. Arrows $X_i \to X_j$ represent the conditional dependency $P(X_j|X_i)$.

In general, parameter independence assumption that every parameter will not be affected by other parameters is made to make the parameter estimation easier. By using this assumption and employing the conjugate prior distribution, the posterior distribution can be represented by

$$P(\Theta_{ij}|\mathcal{D},G) \propto \prod_k \theta_{ijk}^{N'_{ijk}+N_{ijk}-1}.$$

In this equation, N_{ijk} is the number of samples where $X_i = k$ and $\Pi_j = j$, and θ_{ijk} denotes the probability of $X_i = k$ and $\Pi_i = j$ ($\Theta_{ij} = \bigcup_{k=1}^{r_i}\{\theta_{ijk}\}$). For the case of discrete parameter estimation, the Dirichlet distribution is employed as prior distributions. Thus N'_{ijk} represents hyper-parameter of the prior distribution.

The parameter estimation in Bayesian network is carried given a network structure. However, for the case of EDA, Bayesian network for variables is not known in advance. In this case, not only parameter but also network structures have to be estimated from learning data $\mathcal{D}$.

As mentioned in the previous section (Section 5.2.4), EDA generates new individuals from sampling from the predictive posterior distribution $P(\mathbf{X}|\mathcal{D})$. Let G be a DAG and $\mathcal{G}$ be a set of possible network structures. The predictive posterior can be represented by equation (5.23):

$$P(\mathbf{X}|\mathcal{D}) = \sum_{G\in\mathcal{G}}\int d\Theta P(\mathbf{X}|\Theta,G)P(\Theta,G|\mathcal{D}),$$

$$= \sum_{G\in\mathcal{G}}\int d\Theta P(\mathbf{X}|\Theta,G)P(\Theta|\mathcal{D},G)P(G|\mathcal{D}). \qquad (5.23)$$

In general, it is impossible to sum over possible structures. Thus MAP approximation is used to reduce the computational cost (equation (5.24)):

$$P(\mathbf{X}|\mathcal{D}) \simeq \int d\Theta P(\mathbf{X}|\Theta,\widehat{G})P(\Theta|\mathcal{D},\widehat{G}). \qquad (5.24)$$

In this equation, $\widehat{G} \equiv \underset{G\in\mathcal{G}}{\mathrm{argmax}}\, P(G|\mathcal{D})$. This approximation requires calculation of G which gives the maximum posterior probability. By using Bayes theorem,

$$P(G|\mathcal{D}) \propto P(\mathcal{D}|G)P(G).$$

In order to infer the best structure, $P(\mathcal{D}|G)$ is required, where $P(G)$ is a prior distribution. The uniform distribution and heuristics-based prior distribution (e.g, the probabilities of complex networks tend to have smaller probabilities) are often used.

To calculate $P(\mathcal{D}|G)$, [Heckerman *et al.*94] used the Bayes chain rule:

$$P(\mathcal{D}|G) = \prod_{i=1}^{N} P(\mathbf{X}_l | \mathbf{X}_1, \cdots, \mathbf{X}_{l-1}, G).$$

By calculating each term, $P(\mathcal{D}|G)$ is calculated in a following way, which is called BD (Bayesian Dirichlet) metric:

$$P(\mathcal{D}|G) \propto \prod_{i=1}^{n} \prod_{j=1}^{q_i} \frac{\Gamma(N'_{ij})}{\Gamma(N'_{ij} + N_{ij})} \prod_{k=1}^{r_i} \frac{\Gamma(N'_{ijk} + N_{ijk})}{\Gamma(N'_{ijk})}. \tag{5.25}$$

Another way is BIC (Bayesian information criteria) [Schwarz78] approximation (equation (5.26)). BIC approximation gives an approximation of $P(\mathcal{D}|G)$ when $N \to \infty$:

$$\log P(\mathcal{D}|G) \approx \sum_{i=1}^{n} \sum_{j=1}^{q_i} \sum_{k=1}^{r_i} N_{ijk} \log \frac{N_{ijk}}{N_{ij}} - \frac{1}{2} \log N \sum_{i} (r_i - 1) q_i. \tag{5.26}$$

For the case of $N \to \infty$, both BD metric and BIC approximation prefer the same network structure.

5.3.4.4 Problem of Prototype Tree-based Approach

The most of prototype tree-based GP-EDAs adopted the standard parse tree, which is used in a traditional GP. Since the size of the standard parse tree varies in each individual, these GP-EDAs considers a full a_{max}-ary tree (Fig. 5.14(a)), where a_{max} is the maximum arity of the functions. Each node is superposed and the probabilities of each symbol at each position are estimated. For the case of the standard parse tree, we have to consider the case of both function and terminal nodes. For instance, although terminal z and function cos in Fig. 5.14(a) are located at the same depth, these nodes are of a different node type. CPT size in Bayesian network can be calculated by equation (5.27):

$$K = \sum_{i=1}^{n} q_i (r_i - 1). \tag{5.27}$$

In this equation, X_i is an i-th random variable, r_i is the number of possible instances (states) of X_i, Π_i is a set of parent nodes of X_i, and n is the number of variables. For the case of the standard parse tree, $r_i = |\mathfrak{F} \cup \mathfrak{T}|$ and $q_i = |\mathfrak{F} \cup \mathfrak{T}|^{r_i}$. In standard cases, GP uses about 10 symbols. On the other hand, GA uses only 2 symbols. We can see from equation (5.27) that CPT size is greatly affected by the number of symbols. Thus we can easily see that GP-EDAs based on the standard parse tree use much larger CPTs.

In our proposed method, a special chromosome called the expanded parse tree is used in order to overcome the problem of CPT size. In the next section, we describe the expanded parse tree in details.

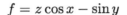

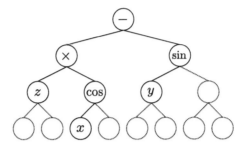

(a) Standard Parse Tree

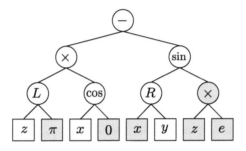

(b) Expanded Parse Tree

FIGURE 5.14: (a) Standard parse tree and (b) EPT, which are syntactically identical. The gray nodes are introns. Circles and square nodes are functions and terminals, respectively.

5.3.4.5   Expanded Parse Tree

In our proposed method, a special chromosome called an expanded parse tree (EPT) is used. EPT was proposed in an algorithm called evolutionary programming with introns (EP-I) [Wineberg and Oppacher94]. EPT is designed to be able to apply GA type genetic operators to tree representations.

Since GA and GP mimic the natural evolution, both algorithms have many features in common. The only difference between two algorithms is their chromosomes. GA adopts linear one-dimensional arrays and crossover is performed by simply exchanging the sub-arrays of the two arrays. GP uses tree structures for chromosomes and interchanges two sub-trees in two trees for crossover. It is well-known that it is possible to convert tree structures into linear arrays. However, the node types (function or terminal) at each index are different among individuals and this difference causes the syntactic destruction after applying GA type crossover (or mutation).

In order to apply GA-type crossover to converted linear arrays (converted from tree structures), node types at each index have to be the same. This condition is satisfied in EPT by inserting introns beneath unused arguments in tree structures. Fig. 5.14(a) is a standard parse tree used in traditional GP and Fig. 5.14(b) is an example of EPT. Note that (a) and (b) are syntactically identical. EPT is expressed by a full a_{max}-ary tree (a_{max} is the maximum arity of the functions). EPT uses extra nodes called selector nodes (L and R) and these nodes are inserted into the standard parse tree. By inserting these selector nodes, all terminal nodes are positioned at the leaves of the full tree. L and R in Fig. 5.14(b) represent selector nodes and these nodes are operated according to $R(x, y) = y$ and $L(x, y) = x$. EPT additionally uses introns in order to ensure that every individual has the same node size. For example in Fig. 5.14, sin and $\times$ have different arity. EPT attaches introns beneath unused arguments (in this case, introns are added under the sin node). After adding introns, sin becomes two-arity function and is operated $\sin(x, y) = \sin(x)$. In Fig. 5.14(b), gray nodes represent introns and these introns do not affect the fitness values of individuals.

In this way, a standard parse tree of maximum arity a_{max} and maximum depth D_P can be converted to a depth D_P full a_{max}-ary EPT with a depth of D_P without changing syntactical meanings. As mentioned above, if the node types at each position are identical among all individuals, GA-type crossover can be applied. Since EPT satisfies this condition, GA-style crossover can be applied to EPT.

Previously, we mentioned that CPT size is very large if we apply Bayesian network to the standard parse tree. For the case of the standard parse tree, we have to consider the possibility of both the function and terminal at each position. On the other hand, for the case of EPT, we only have to consider the possibility of functions at branches and terminals at leaves.

The size of CPT is greatly affected by the number of possible symbols at each position (equation (5.27)). As a consequence, by using EPT, the size of CPT can be greatly reduced.

5.3.4.6 Estimation of Bayesian Network

As mentioned in Section 5.2.4, new individuals in EDA are generated according to the predictive posterior represented below:

$$P(\mathbf{X}|\mathcal{D}_g) = \sum_G \int d\Theta \, P(\mathbf{X}|G, \Theta) P(\Theta|\mathcal{D}_g, G) P(G|\mathcal{D}_g).$$

In this equation, $\mathbf{X} = \bigcup\{X_i\}$ is an individual, $\mathcal{D}_g = \bigcup\{\mathbf{X}_i\}$ is a set of promising solutions at generation g, and G is a directed acyclic graph (DAG). Since it is impossible to sum over possible G, only one G which yields the maximum posterior probability is used. That is

$$\hat{G} = \underset{G}{\operatorname{argmax}} P(G|\mathcal{D}_g).$$

By using this approximation, new individuals are generated using the following equation.

$$P(\mathbf{X}|\mathcal{D}_g) = \int d\Theta\, P(\mathbf{X}|\widehat{G},\Theta)P(\Theta|\mathcal{D}_g,\widehat{G})$$

$$= \prod_{i=1}^{n}\prod_{j=1}^{q_i}\frac{N_{ijk} + N'_{ijk}}{N_{ij} + N'_{ij}} \qquad (5.28)$$

In this equation, n is the number of nodes. N_{ijk} is the number of samples where i-th node is at k-th state and Π_i (parent of i-th node) is at j-th state. $N_{ij} = \sum N_{ijk}$. N'_{ijk} are hyper-parameters of prior distributions and we set $N'_{ijk} = 1$ (uniform).

For inference of $\widehat{G}$, which yields the maximum posterior probability, many algorithms have been proposed. In our algorithm, the combination of BIC approximation and K2 algorithm [Cooper and Herskovits92] are used.

Although the K2 algorithm can quickly infer the network with the maximum posterior probability, because the GP uses many nodes during the search, it requires many computational power even with K2 algorithm. In tree structures, neighboring nodes exhibit stronger dependencies than distant nodes. We use this assumption for a faster network inference.

Let $U(X_i, R_P)$ be a node positioned at R_P levels above X_i. Nodes satisfying the following two conditions are candidates to be parents of X_i in POLE.

- Nodes that belong to the sub-tree rooted at $U(X_i, R_P)$.

- Nodes whose indices are smaller than i.

We describe an example of the parent candidates of X_i in Fig. 5.15. In this figure, gray nodes are parent candidates of X_9 ($R_P = 2$).

5.3.4.7 Generation of New Individuals

New individuals are generated according to equation (5.28). As can been seen with this equation, probabilistic logic sampling (PLS) can be used to sample this equation. PLS is a highly effective sampling method and sampling cost of PLS is very small. Procedures of PLS can be summarized below.

- Nodes that have no parent nodes are determined using the probability distribution

- Nodes whose parents are already decided are determined using the probability distribution

Since Bayesian network is DAG, all nodes can be decided using PLS. Fig. 5.16(b) shows an example of the node decision process corresponding to the network shown in Fig. 5.16(a).

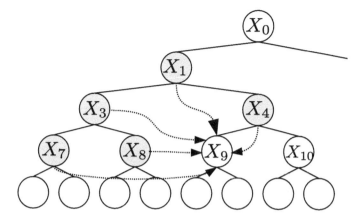

FIGURE 5.15: Gray nodes represent parent candidacies for X_9 ($R_P = 2$).

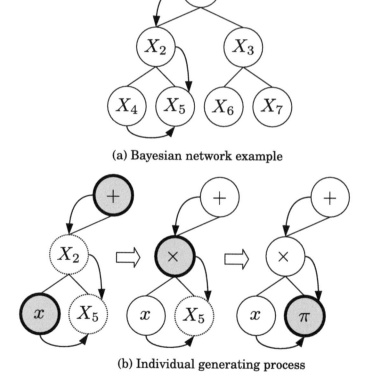

(a) Bayesian network example

(b) Individual generating process

FIGURE 5.16: Bayesian network example (a) and individual generation process (b).

Table 5.2: Main parameters of POLE.

Symbol	Meaning	Value
M	Population size	—
G	Generation to run	—
D_P	Maximum tree depth	—
P_s	Selection rate	0.1 (MAX problem) 0.1 (DMAX problem) 0.2 (Royal Tree problem)
R_P	Parent range	2 (MAX problem) 2 (DMAX problem) 4 (Royal Tree problem)
k	The number of max incoming edge	∞
P_e	Elite rate	0.005
P_F	Function selection rate at the initialization	0.8

5.3.4.8 Comparative Experiment

We applied POLE to three distinct benchmark tests. The main parameters of POLE are listed in Table 5.2.

In comparative experiments, we compared the performance of the four models listed below.

- **POLE**
  This is our proposed method.

- **Univariate model (PIPE model)**
  This algorithm is a univariate case of POLE. No networks are constructed in this algorithm.

- **Adjacent model (EDP model)**
  This model constructs parent and child relationships in the tree structure according to EDP.

- **Simple GP (GP)**
  This algorithm is a traditional GP. In the experiments, $P_e = 0.005$ (elite rate), $P_c = 0.995$ (crossover rate), $P_m = 0$ (mutation rate), and $P_r = 0$ (reproduction rate) are used. The tournament size is set to $t_s = 2$.

Since the best population size is different in each algorithm, we employed following procedures to decide the population size.

Every algorithm starts from $M = 100$, and increase the population size ($M = 100, 160, 250, 400, 630, 1000, 1600, \cdots$, max $M = 25000$). At each population size, 20 runs are carried. If the optimum is obtained 20 times from

Table 5.3: The number of fitness evaluations for the MAX problem. Values in parentheses represent the standard deviation (stdev) for each case.

		$D_P = 5$	$D_P = 6$	$D_P = 7$
POLE	Average	487	1,758	5,588
	Stdev	(21)	(59)	(75)
Univariate	Average	497	1,786	5,688
	Stdev	(21)	(108)	(46)
Adjacent	Average	581	2,496	7,401
	Stdev	(28)	(186)	(213)
GP	Average	1,209	3,962	11,141
	Stdev	(268)	(825)	(3,906)

Table 5.4: t-test for the MAX problem. Each value represents p-values.

t-value	$D_P = 5$	$D_P = 6$	$D_P = 7$
POLE vs. Univariate	3.30×10^{-1}	4.86×10^{-1}	2.69×10^{-3}
POLE vs. Adjacent	1.85×10^{-7}	1.47×10^{-7}	3.06×10^{-11}
POLE vs. GP	1.27×10^{-5}	1.37×10^{-5}	1.50×10^{-3}

the 20 runs, the algorithm stops increasing and the average number of fitness evaluations F_{avr} is calculated. Since F_{avr} tends to have a large standard deviation, F_{avr} is calculated 10 times, and conduct a t-test using the average of F_{avr}.

5.3.4.8.1 Experiment 1: MAX problem We first applied POLE to the MAX problem (see Section 5.2.6).

Table 5.3 shows the number of fitness evaluations along with the standard deviations. We plotted these values in Fig. 5.17, in which vertical axis is the number of fitness evaluations and horizontal axis is the tree size. Tree size stands for the number of nodes contained in the optimum structure. We can see from Fig. 5.17 that GP requires more fitness evaluations to obtain the optimum. Although POLE shows the best performance among four methods, the number of fitness evaluations is indistinguishable from that of the Univariate model. Table 5.4 is the results of t-test. We can see that p-value for POLE and Univariate model is smaller than 1%. This indicates that the average number of fitness evaluations of POLE and Univariate model is statistically different. Because the MAX problem does not have deceptiveness, there are no dependencies between nodes. Thus the search performances of POLE and the Univariate model are almost identical. Although the search performance of the Adjacent model is better than GP, it is worse than POLE and the Univariate model. Since there are no interactions among nodes in the MAX problem, unnecessary dependency estimation degraded the search

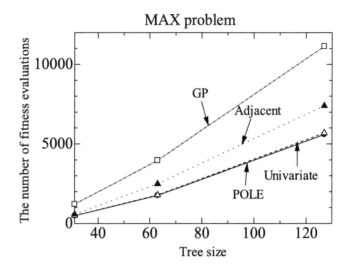

FIGURE 5.17: The number of fitness evaluations for the MAX problem at each problem size.

performance.

5.3.4.8.2 Experiment 2: DMAX problem We applied POLE to the DMAX problem (see Section 5.2.6), which is a highly deceptive benchmark test for the traditional GP.

Table 5.5 represents the number of fitness evaluations along with the standard deviations. We plotted these values in Fig. 5.18. Since the Univariate and Adjacent model could not obtain the optimum of the DMAX problem even at once, we did not show the results of these methods. Table 5.6 is the results of t-test. We can see that for $D_P = 4$, p-value for POLE and GP is extremely small, indicating the statistical difference between these two methods.

As can been seen with these results, we can see that GP requires 10 times more fitness evaluations compared to POLE. GP cannot solve the DMAX problem because of the deceptive fitness landscape of the problem.

Fig. 5.19(d) describes the global optimum of the DMAX problem. (a) is a local optimum and (b) and (c) are intermediate structures which are similar to the global and a local optimum. As can been seen, although a fitness difference of (a) and (d) is relatively small, the structures between (a) and (d) are not similar. On the other hand, structural similarity between (d) and (c) is large, but the fitness difference between these two structures is large. As mentioned in the previous section, GP-type sampling is effective in the case that two structurally similar individuals have similar fitness values. The

Table 5.5: The number of fitness evaluations for the DMAX problem. Values in parentheses represent the standard deviation (stdev) for each case.

		$D_P = 3$	$D_P = 4$
POLE	Average	1,570	124,531
	Stdev	(155)	(23,403)
Univariate	Average	1,409	−
	Stdev	(107)	−
Adjacent	Average	1,440	−
	Stdev	(119)	−
GP	Average	11,105	1,632,159
	Stdev	(6,958)	(645,696)

Table 5.6: t-test for the DMAX problem. Each value represents p-values.

t-value	$D_P = 3$	$D_P = 4$
POLE vs. Univariate	1.58×10^{-2}	−
POLE vs. Adjacent	5.10×10^{-2}	−
POLE vs. GP	1.89×10^{-3}	4.14×10^{-5}

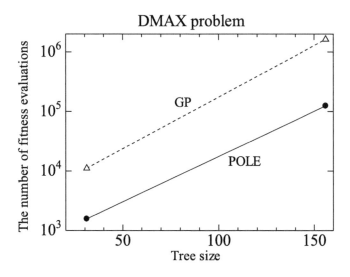

FIGURE 5.18: The number of fitness evaluations for the DMAX problem at each problem size.

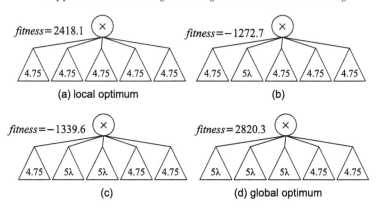

FIGURE 5.19: The local (a) and global optima (d) for the DMAX problem ($m = 5$, $r = 3$, $D_P = 3$). (b) and (c) represent intermediate structures between (a) and (d).

DMAX problem does not satisfy the condition of GP-sampling at all, and as a consequence, GP showed very poor performance compared to that of POLE.

In this problem, there are two types of building blocks: $\lambda + \lambda + \lambda + \lambda + \lambda$ and $0.95 + 0.95 + 0.95 + 0.95 + 0.95$. Since the Univariate and the Adjacent models cannot estimate the dependencies, these algorithms often destroy the building blocks. POLE constructs the Bayesian network and can grasp the dependencies exhibited in the DMAX problem. As a result, POLE effectively obtained the optimum with the fewest number of fitness evaluations. We can say that POLE is highly effective for solving the DMAX problem in comparison with other program evolution methods.

We describe an estimated network structure by POLE in Fig. 5.20. This network is a network at the last generation in a randomly picked up trial ($M = 4000$, $D_P = 4$). We can see from this fiure that POLE successfully estimated the interactions between nodes. As mentioned in the Section 5.2.6, this problem has two types of building blocks. POLE estimated these two types of building blocks. Since POLE can identify these two interactions effectively, POLE obtained the optimum solutions with the lower number of fitness evaluations.

5.3.4.8.3 Experiment 3: Royal tree problem   We applied POLE to the royal tree problem. The royal tree problem used in this section is slightly different from that of the original one. The original royal tree problem only uses a terminal x. The authors of the royal tree problem have pointed out that the deceptiveness can be added to this problem by offering intermediate scores to terminals y, z, etc. In this section, we set $Score(x) = 1$, $Score(y) = 0.95$ and also define a perfect tree using y. Furthermore, we assumed that all the functions have an arity of 2. Fig. 5.21 shows a scoring system for the royal

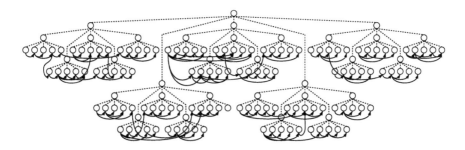

FIGURE 5.20: An estimated network for the DMAX problem ($M = 4,000$, $D_P = 4$).

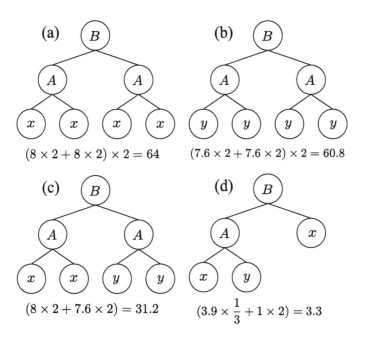

FIGURE 5.21: The scoring system for the royal tree problem with deceptiveness.

Table 5.7: The number of fitness evaluations for the royal tree problems. Values in parentheses represent the standard deviation (stdev) for each case.

		$D_P = 4$	$D_P - 5$	$D_P = 6$	$D_P = 7$
POLE	Average	$1,357$	$7,711$	$38,039$	$147,030$
	Stdev	(339)	$(1,291)$	$(5,847)$	$(13,191)$
Univariate	Average	$4,135$	$98,837$	$-$	$-$
	Stdev	(864)	$(29,131)$	$-$	$-$
Adjacent	Average	$5,752$	$90,080$	$-$	$-$
	Stdev	$(1,791)$	$(20,016)$	$-$	$-$
GP	Average	$2,331$	$35,238$	$-$	$-$
	Stdev	(989)	$(16,234)$	$-$	$-$

Table 5.8: t-test for the royal tree problem. Each value represents p-values.

t-value	$D_P = 4$	$D_P = 5$
POLE vs. Univariate	7.89×10^{-7}	3.83×10^{-6}
POLE vs. Adjacent	2.20×10^{-5}	3.62×10^{-7}
POLE vs. GP	1.32×10^{-2}	4.46×10^{-4}

tree problem used in this section.

Table 5.7 shows the number of fitness evaluations and its standard deviations. Fig. 5.22 describes these values. Since GP, the Univariate model, and the Adjacent model could not obtain the optimum at some D_P, the results of these methods are partially omitted. We can see from these results that POLE obtained the optimum of the royal tree problem with the fewest number of fitness evaluations. For the case of $D_P = 6$ and $D_P = 7$, only POLE succeeded to obtain the optimum with a probability of 100%. Table 5.8 is the results of t-test for $D_P = 4$ and 5. We can see that almost all p-values for POLE and the other methods are smaller than 1%.

In the DMAX problem, there are horizontal dependencies (dependencies between sibling nodes). On the other hand, the royal tree problem has two types of interactions: parent child interactions (e.g, children of function E node must be function D node) and horizontal interactions (e.g, siblings of x must be x). Since the Univariate and Adjacent models could not estimate these interactions, this method failed to obtain the optimum.

GP succeeded to obtain the optimum with relatively higher probability. GP sometimes trapped at a local optimum which is composed of terminal y. For the case of GP, it is almost impossible to escape from a local optimum once trapped. Because this problem also violates the assumption required for GP-type sampling, GP showed poorer performance compared to POLE.

We show estimated network structures by POLE in Fig. 5.23(a) and (b). A network (a) is a network at earlier generation (generation 2) and (b) is

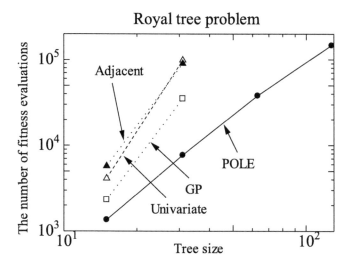

FIGURE 5.22: The number of fitness evaluations for the royal tree problem at each problem size.

at the later generation (generation 10). These networks are chosen from a randomly selected trial ($M = 4000$, $D_P = 6$). From Fig. 5.23(a), it can be seen that POLE estimates the interactions exhibited between parents and children nodes. The optimum of the royal tree problem has a vertical ordering $D, C, B, \cdots$, and POLE estimates these interactions in earlier generations. On the other hand, Fig. 5.23(b) shows the network constructed among sibling nodes. POLE estimates these interactions in the later generations. Since POLE grasps these interactions, POLE can escape from the local optima.

5.3.4.8.4 Expanded Parse Tree vs. Standard Parse Tree In this section, we experimentally show the effectiveness of EPT against the standard parse tree (SPT). In order to confirm the effectiveness, we repeated the experiment using standard parse tree, keeping network models and other parameters unchanged. We used three network models: Bayesian network, the Univariate, and the Adjacent network (these three methods are referred to as Bayesnet + SPT, univariate + SPT, and adjacent + SPT respectively, in Table 5.9). Table 5.9 represents the number of fitness evaluations in six methods along with the standard deviations. For the case of Bayesian network, EPT is superior in almost every setting. Especially, for the case of $D_P = 6, 7$, only Bayesian network with EPT obtained the optimum of the royal tree problem. Table 5.10 is the results of t-test for $D_P = 4$ and 5 between two types of chromosomes. We can see that p-value between EPT and SPT for Bayesian network is the smallest among other network structures. In tree

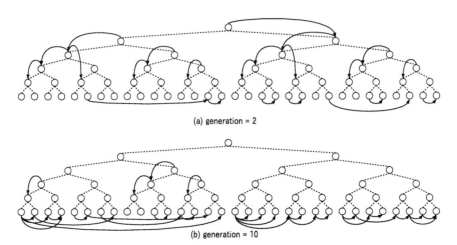

(a) generation = 2

(b) generation = 10

FIGURE 5.23: Estimated networks for the royal tree problem ($M = 4,000$, $D_P = 6$).

Table 5.9: The number of evaluations in two types of chromosomes (EPT: expanded parse tree and SPT: standard parse tree) and their corresponding standard deviations.

		$D_P = 4$	$D_P = 5$	$D_P = 6$	$D_P = 7$
Bayesnet + EPT	Average	1,357	7,711	38,039	147,030
	Stdev	(339)	(1,291)	(5,847)	(13,191)
Bayesnet+ SPT	Average	1,429	11,003	−	−
	Stdev	(362)	(2,554)	−	−
Univariate + EPT	Average	4,135	98,837	−	−
	Stdev	(864)	(29,131)	−	−
Univariate + SPT	Average	4584	117432	−	−
	Stdev	(1,338)	(38,959)	−	−
Adjacent + EPT	Average	5,752	90,080	−	−
	Stdev	(1,791)	(20,016)	−	−
Adjacent + SPT	Average	7,765	112,494	−	−
	Stdev	(2,220)	(34,258)	−	−

Table 5.10: t-test for the royal tree problem in two types of chromosomes. Each value represents p-values.

	$D_P = 4$	$D_P = 5$
Bayesnet EPT vs. SPT	6.51×10^{-1}	2.90×10^{-3}
Univariate EPT vs. SPT	3.86×10^{-1}	2.44×10^{-1}
Adjacent EPT vs. SPT	3.92×10^{-2}	8.28×10^{-2}

structures with larger depth, they contain more noises. It is more difficult to construct Bayesian networks using contaminated data. In the univariate approach, we cannot see a difference in the search performance between EPT and the standard parse tree. Because the univariate approach does not construct network structure, it is less vulnerable to noisy data. Similar to the case of the univariate approach, the adjacent approach also does not construct network from learning samples. Although EPT-based methods showed the better performance than adjacent approach, this difference is smaller than what was seen in the case of Bayesian network.

5.3.4.9 Discussion

In this section, we introduced POLE, which employed a Bayesian network for estimation of interactions between nodes. We empirically showed that our proposed method is highly effective for the problems which exhibit strong dependencies between nodes. Because the Univariate and the Adjacent models do not estimate the interactions between nodes, these algorithms cannot reuse the building blocks.

We should discuss the overhead of the inference of Bayesian network in POLE method. Since POLE is accompanied by a computationally expensive task, POLE requires more computational power than the Univariate or the Adjacent model which does not estimate the interactions. However, since we adopted the assumption that there are few interactions among distant nodes, POLE dramatically reduces the computational time devoted to network construction.

5.3.5 Summary of Prototype Tree-based Approach

In this section, we have introduced GP-EDAs based on prototype trees. Since prototype tree-based GP-EDA is an extension of GA-EDA, techniques introduced in GA-EDA are used. PIPE and ECGP extended PBIL and ECGA, respectively. POLE is also an extension of BOA and EBNA, which uses Bayesian networks. In GA-EDA, more advanced graphical models have been proposed [Santana05, Pena *et al*.05]. Especially when using EPT, these advanced graphical models can be imported with almost no modifications. GA-EDAs employing these graphical models are reported to be more powerful than those based on a simple Bayesian network.

Furthermore, a hybrid approach using crossover and mutation may overcome the drawbacks of GP-EDA approaches. Yanai and Iba combined EDP approach with traditional GP [Yanai and Iba04]. Since general prototype tree-based GP-EDAs can not estimate the building blocks which are position independent, it is considered that this extension overcomes the drawbacks of prototype tree-based GP-EDAs. Actually, they have reported that combination of the EDP and GP approaches are highly effective for improving the search performance [Yanai and Iba04].

PPT-based method sometimes suffers from problems arising from function nodes having different number of child nodes. These function nodes cause intron nodes, which do not affect the fitness function. Moreover, the function nodes having many child nodes increase the search space and the number of samples necessary for properly constructing the probabilistic model. In order to solve this problem, Yanase and Iba have proposed binary encoding for PPT. They convert each function node to a subtree of binary nodes where the converted tree is correct in grammar. Their method reduces ineffectual search space, and the binary encoded tree is able to express the same tree structures as the original method (see [Yanase and Iba09] for details).

5.4   PCFG-based Methods

As mentioned in the previous section, prototype tree-based approaches can easily employ techniques introduced in GA-EDAs. However, because prototype trees are fixed length structures and are limited to an a_{max} full tree structure (a_{max} is the largest arity among functions), the number of nodes is exponential to a_{max}. This gives rise to the increase of computational cost for estimation of the interactions. Furthermore, prototype tree-based GP-EDAs are basically position-dependent methods and are not suitable for estimating position-independent building blocks. PCFG-based GP-EDAs suffer from neither of the problems above and, as a consequence, many PCFG-based GP-EDAs have been proposed. The advantages of PCFG-based GP-EDAs are listed below.

- Chromosomes of PCFG-based GP-EDA need not be fixed length

- PCFG-based methods can estimate position independent building blocks

For the cases of prototype tree-based methods, it is relatively easy to estimate the interactions among nodes. On the other hand, because of the context freedom assumption in PCFG, PCFG-based methods have difficulty in estimating the interactions among nodes. In the next section, we introduce basics of PCFG for understanding of PCFG-based GP-EDAs.

5.4.1   PCFG

Because all GP-EDAs, which will be explained in the following sections, are based on PCFG, we explain basic concepts of PCFG in this section.

We first define a context-free grammar (CFG). CFG is defined by four variables $G = \{\mathcal{N}, \mathcal{T}, \mathcal{R}, \mathcal{B}\}$, where G is a CFG. The meaning of these variables are listed below.

- $\mathcal{N}$: A finite set of non-terminal symbols

- $\mathcal{T}$: A finite set of terminal symbols

- $\mathcal{R}$: A finite set of production rules

- $\mathcal{B}$: Start symbol

Note that the terms "non-terminal" and "terminal" in CFG are different from those in GP (for example in regression problem, not only x, y but also sin, + are treated as terminals in CFG).

Production rules in CFG can be expressed as follows:

$$A \rightarrow \alpha \ (A \in \mathcal{N}, \alpha \in (\mathcal{N} \cup \mathcal{T})^*) \tag{5.29}$$

In equation (5.29), $(\mathcal{N} \cup \mathcal{T})^*$ represents a set of possible sentences composed of $(\mathcal{N} \cup \mathcal{T})$. By applying production rules to the start symbol $\mathcal{B}$, grammar G generates sentences. A language generated by grammar G is represented by $L(G)$.

By applying production rules, a non-terminal A is replaced by other symbols. For example, if we apply a production rule represented by equation (5.29) to $\alpha_1 A \alpha_2 (\alpha_1, \alpha_2 \in (\mathcal{N} \cup \mathcal{T})^*, A \in \mathcal{N})$, this string becomes $\alpha_1 \alpha \alpha_2$. $\alpha_1 A \alpha_2$ derived $\alpha_1 \alpha \alpha_2$ and this process is represented as follows:

$$\alpha_1 A \alpha_2 \underset{G}{\Rightarrow} \alpha_1 \alpha \alpha_2.$$

Furthermore, if

$$\alpha_1 \underset{G}{\Rightarrow} \alpha_2 \cdots \underset{G}{\Rightarrow} \alpha_n (\alpha_i \in (\mathcal{N} \cup \mathcal{T})^*),$$

it is said that α_n is derived from α_1, and is represented by $\alpha_1 \underset{G}{\overset{*}{\Rightarrow}} \alpha_n$. This derivation process can be expressed by tree structures, and these trees are called the derivation trees. Derivation trees of grammar G are defined as follows.

1. A node is an element of $(\mathcal{N} \cup \mathcal{T})$

2. A root is $\mathcal{B}$

3. A branch node is an element of $\mathcal{N}$

4. If children of $A \in \mathcal{N}$ are $\alpha_1 \alpha_2 \cdots \alpha_k \ (\alpha_i \in (\mathcal{N} \cup \mathcal{T}))$ from left, production rule $A \rightarrow \alpha_1 \alpha_2 \cdots \alpha_k$ is an element of $\mathcal{R}$

Let us explain CFG with an intuitive example. Suppose we are considering general univariate functions composed of sin, cos, exp, log, and arithmetic operators. A grammar $G_{regression}$ can be represented as follows. In this case, start symbol is $\langle expr \rangle$. A set of non-terminals is

$$\mathcal{N} = \{\langle expr \rangle, \langle op2 \rangle, \langle op1 \rangle, \langle var \rangle, \langle const \rangle\},$$

and for terminals,

$$\mathcal{T} = \{+, -, \times, \div, \sin, \cos, \exp, \log, x, C\}.$$

One possible production rules can be represented by a table below.

#	Production rule
0	$\langle expr \rangle \rightarrow \langle op2 \rangle \langle expr \rangle \langle expr \rangle$
1	$\langle expr \rangle \rightarrow \langle op1 \rangle \langle expr \rangle$
2	$\langle expr \rangle \rightarrow \langle var \rangle$
3	$\langle expr \rangle \rightarrow \langle const \rangle$
4	$\langle op2 \rangle \rightarrow +$
5	$\langle op2 \rangle \rightarrow -$
6	$\langle op2 \rangle \rightarrow \times$
7	$\langle op2 \rangle \rightarrow \div$
8	$\langle op1 \rangle \rightarrow \sin$
9	$\langle op1 \rangle \rightarrow \cos$
10	$\langle op1 \rangle \rightarrow \exp$
11	$\langle op1 \rangle \rightarrow \log$
12	$\langle var \rangle \rightarrow x$
13	$\langle const \rangle \rightarrow C$ (constant)

A grammar defined above is a language for generating univariate functions. For example, $G_{regression}$ generates the following function.

$$
\begin{aligned}
\langle expr \rangle &\rightarrow \langle op2 \rangle \langle expr \rangle \langle expr \rangle \\
&\rightarrow + \langle expr \rangle \langle expr \rangle \\
&\rightarrow + \langle op2 \rangle \langle expr \rangle \langle expr \rangle \langle expr \rangle \\
&\rightarrow + + \langle expr \rangle \langle expr \rangle \langle expr \rangle \\
&\rightarrow + + \langle op1 \rangle \langle expr \rangle \langle expr \rangle \langle expr \rangle \\
&\rightarrow + + \log \langle expr \rangle \langle expr \rangle \langle expr \rangle \\
&\rightarrow + + \log \langle var \rangle \langle expr \rangle \langle expr \rangle \\
&\rightarrow + + \log x \ \langle expr \rangle \langle expr \rangle \\
&\rightarrow + + \log x \ \langle var \rangle \langle expr \rangle \\
&\rightarrow + + \log x \, x \ \langle expr \rangle \\
&\rightarrow + + \log x \, x \ \langle const \rangle \\
&\rightarrow + + \log x \, x \, C
\end{aligned}
$$

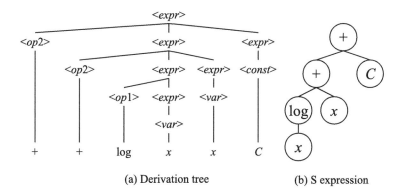

(a) Derivation tree   (b) S expression

FIGURE 5.24: A derivation tree for $\log x + x + C$ and its corresponding tree structure in GP.

In this case, the derived function is

$$f(x) = \log x + x + C.$$

The derivation process above can be represented by a derivation tree in Fig. 5.24(a).

In general GP, these functions are represented by standard tree representations (Fig. 5.24(b)). We can see that derivation trees also have ability to express the same functions. GGGP (grammar guided GP) [Whigham95, Whigham96] took advantage of this idea, and this algorithm adopted derivation trees for its chromosome. In GGGP, functions and programs derived from derivation trees correspond to P-TYPE. Genetic operators of GGGP are identical to those of standard GP. GGGP also selects promising solutions and these solutions participate in mutation or crossover. In standard GP, crossover points are selected randomly and the crossover is done by swapping randomly selected two sub-trees. On the other hand, GGGP only swaps between the same non-terminal nodes. This limitation in GGGP is the same as that of typed GP.

PCFG adds probabilities to each production rules in CFG. For example, we show the probabilty examples of production rules for $\# 0 \sim 3$ below:

$$P(\langle expr \rangle \rightarrow \langle op2 \rangle \langle expr \rangle \langle expr \rangle) = 0.4$$
$$P(\langle expr \rangle \rightarrow \langle op1 \rangle \langle expr \rangle) = 0.3$$
$$P(\langle expr \rangle \rightarrow \langle var \rangle) = 0.1$$
$$P(\langle expr \rangle \rightarrow \langle const \rangle) = 0.2.$$

Because the probabilities of the production rules are given condional to the left-hand side of the rules, the probabilities are normalized in terms of $\langle expr \rangle$

Table 5.11: Comparison of PCFG-based GP-EDA.

	Grammar	Position Independence	Adaptive Model Learning
SG-GP	PCFG	Yes	No
Vectorial SG-GP	PCFG with depth	No	No
PEEL	L-system	No	Yes
GMPE	Model Merging PCFG	Yes	Yes
GT-EDA	Tree Ajoining Grammar	No	Yes
BAP	PCFG with Bayesnet	No	Yes
PAGE	PCFG-LA	Yes	Yes

in the example:

$$\sum_{\alpha} P(\langle expr \rangle \rightarrow \alpha) = 1.$$

In PCFG, likelihood functions of derivation trees are calculated by multiplying the probabilities of production rules, which appear in the target derivation trees. For a notational convenience, let $P(\#1)$ be a probability of production rule of $\#1$ in a previous table, for instance $(P(\#1) \equiv P(\langle expr \rangle \rightarrow \langle op1 \rangle \langle expr \rangle))$. A likelihood of the derivation tree represented by Fig. 5.24(a) is calculated as follows:

$$P(\#0)^2 P(\#1) P(\#2)^3 P(\#4)^2 P(\#11) P(\#12)^2 P(\#13).$$

One of the most important task in PCFG is to infer the probabilities (parameters) of production rules. Generally, derivation trees are not known. Consequently, the parameter inference can not be carried out by just counting the frequencies of rules. In PCFG, more advanced parameter inference methods as the expectation maximization (EM) algorithm or variational Bayes is used.

We next review PCFG-based GP-EDAs in the following sections. Table 5.11 depicts features of each PCFG-based GP-EDAs.

5.4.2   SG-GP

SG-GP (stochastic grammar-based GP) [Ratle and Sebag01] is the simplest PCFG-based GP-EDA. SG-GP adopted the basic PCFG explained in the previous section. In NLP, parameters of the PCFG are inferred without observing the derivation trees by using EM (expectation maximization) algorithm. On the other hand, for the case of GP-EDAs, parameters can be easily estimated by counting the frequencies of corresponding rules because derivation trees are observed.

In SG-GP, all the rules r_i are attached weight w_i. When generating new individuals, a rule r_i is selected proportional to w_i. The selection of rule r_i is done with probability p_i:

$$p_i = \frac{w_i}{\sum w_j}.$$

For the case of NLP, the parameters (probabilities) of rules are estimated by maximum likelihood. SG-GP employs the incremental strategy inspired by PBIL. If a rule r_i appears in b individuals among N_b best individuals, the weight w_i is magnified by $(1 + \epsilon)^b$ times, where ϵ is the learning rate. On the other hand, if r_i is carried by w individuals among N_w worst individuals, w_i is divided by $(1 + \epsilon)^b$.

The PCFG model adopted in SG-GP strongly assumes the context freedom assumption that the probabilities of production rules do not depend on the context they appear. As a consequence, SG-GP basically cannot estimate building blocks since they exhibit strong dependencies among nodes. Furthermore, because the probabilities of large tree structures tend to be very small, SG-GP is not suitable for the problems requiring many nodes. The authors of SG-GP have also proposed an extended algorithm named vectorial SG-GP, which takes into account the depth information. This extension weakens the context freedom assumption. However, at the same time, this extension also loses the merit of position independence of PCFG because probabilities of production rules in vectorial SG-GP heavily depend on their depth. This depth dependence is employed in later GP-EDAs as PEEL and GT-EDA.

5.4.3   PEEL

PEEL (program evolution with explicit learning) [Shan *et al.*03] employed the SSDT (search space description table), which was inspired by stochastic parametric L-system (Lindenmayer system). L-system is identical to PCFG with two additional parameters. PEEL adopted the production rules represented by equation (5.30):

$$X(d, \ell) \to \zeta. \tag{5.30}$$

In this equation, d is a depth at which the production rule appears and ℓ represents a relative position. PEEL starts from general grammars, where "don't care" symbol # is used. For example, production rules with $d = \#$ or $\ell = \#$ do not care the depth or the position, respectively. PEEL gradually specializes the grammar towards the optimum structure.

For expressing the importance of the production rules, "pheromone" is used. Pheromone corresponds to probability tables in GP-EDA, and is updated using ant colony optimization.

5.4.4   GMPE

GMPE (grammar model-based program evolution) [Shan *et al.*04] is inspired by the algorithm called the model merging method [Stolcke94]. GMPE employs a very flexible grammar that is learnt from promising solutions, and is capable of estimating the interactions between nodes.

GMPE starts from the production rules that exclusively generate learning data (promising solutions). GMPE gradually merges two non-terminal symbols by making the production rules more general. Let V, W, and ζ_i be $V, W \in \mathcal{N}$ and $\zeta_i \in (\mathcal{N} \cup \mathcal{T})^*$. Suppose there are two production rules:

$$V_1 \rightarrow \zeta_1, V_2 \rightarrow \zeta_2.$$

If GMPE decides to merge V_1 and V_2, with $merge(V_1, V_2) = W$, these two rules are converted into a following rule:

$$W \rightarrow \zeta_1, W \rightarrow \zeta_2.$$

After this conversion, resulting rules become more general. GMPE employs the MDL principle for the decision of merger. GMPE tries to find a grammar which is relatively simple but can cover learning data well. In general, these two factors are in a relationship of trade off. MDL incorporates these two factors. MDL can be represented by equation (5.31):

$$L_C(D) = L_C(G) + L_C(\mathcal{D}|G). \tag{5.31}$$

For the case of GMPE, G is grammar and $\mathcal{D}$ is training data. $L_C(Z)$ is a cost of coding Z. $L_C(\mathcal{D}|G)$ corresponds to a likelihood of $\mathcal{D}$. $L_C(G)$ is a model complexity and represents the coding length of a grammar G.

5.4.5   Grammar Transformation in an EDA

Grammar transformation in an EDA (GT-EDA) [de Jong and Bosman04] focuses on extracting the building blocks in derivation trees. GT-EDA employs expanded production rules to consider the interactions. Although the underlying concepts of GT-EDA and GMPE are the same, methods of obtaining rules (grammars) are different. As explained above, GMPE infers the production rules from the most specific grammar. On the other hand, GT-EDA estimates a grammar from the most general one.

For simplicity, we consider the symbolic regression problem using $\sin, +$ and x. Corresponding production rules can be represented by the following equations:

$$\mathcal{S} \rightarrow \sin \mathcal{S},$$
$$\mathcal{S} \rightarrow +\mathcal{S}\,\mathcal{S},$$
$$\mathcal{S} \rightarrow x.$$

These rules generate general functions composed of $\{\sin, +, x\}$. As mentioned previously, these production rules cannot consider contexts because $\mathcal{S}$ deriving $\sin$ and $\mathcal{S}$ deriving $+$ cannot be distinguished in these equations. In order to overcome this drawback, GT-EDA instead uses expanded production rules represented below:

$$\mathcal{S} \to \sin \mathcal{S},$$
$$\mathcal{S} \to +\mathcal{S}\mathcal{S},$$
$$\mathcal{S} \to +(\sin \mathcal{S})\mathcal{S},$$
$$\mathcal{S} \to x.$$

In these expanded rules, a production rule $\mathcal{S} \to +\mathcal{S}\mathcal{S}$ is specialized by specifying one $\mathcal{S}$ in the right side of the rule. This expansion corresponds to considering the interactions between nodes. In this case, this expanded production rule means considering the interaction between $+$ and $\sin$.

Since the expansion of the production rules increases the number of rules exponentially, GT-EDA expands the rules only once. GT-EDA also uses the MDL principle to decide which rules to expand.

5.4.6 PAGE

In the previous sections, we gave a brief explanation of several PCFG-based GP-EDAs. We next show a detailed explanation for a state-of-art prototype tree PCFG-based GP-EDA named PAGE (programming with annotated grammar estimation) [Hasegawa and Iba09]. PAGE employed PCFG-LA (PCFG with latent annotations) for a statistical model. The results of this section are taken from [Hasegawa and Iba09].

5.4.6.1 Introduction

In this section, we introduce GP-EDA based on PCFG with latent annotations [Matsuzaki *et al.*05].

The conventional PCFG takes advantage of the context freedom assumption that the probabilities of the rules are not dependent on their contexts (their parents or sibling nodes, for example). GP-EDA based on the conventional PCFG, SG-GP for instance, estimates parameters of production rules and generates programs by sampling from the estimated distributions. Although the use of the context freedom assumption eases the statistical estimation, this is at the expense of losing the information about dependencies between nodes. Since building blocks in GP often exhibit strong dependencies between nodes, GP-EDA based on the conventional PCFG cannot estimate these building blocks. In NLP, many approaches have been considered to weaken the context freedom assumption. Recent GP-EDAs, such as GMPE (Section 5.4.4) and GT-EDA (Section 5.4.5), adopt more advanced PCFGs. These methods

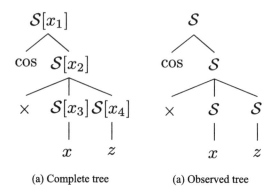

(a) Complete tree   (a) Observed tree

FIGURE 5.25: A complete tree with annotations (a) and its observed tree (b).

infer the production rules themselves from learning data. A different approach which can encode dependencies is to use annotations. In NLP, many types of annotations have been proposed up to now and these annotations have already been applied to GP-EDAs. Tanev proposed a PCFG-based GP which incorporates annotations in PCFG [Tanev04, Tanev05]. His approach employs parent and sibling nodes for context information (the probabilities of production rules depend on their parent or sibling nodes). GT-EDA uses depth information (depth where production rules appear in derivation trees) and this information can be considered as depth annotations. Adding these annotations does not increase the computational cost at all and these fixed annotations are effective in some extent. However, when using these fixed annotations, probabilities of production rules depend on parents, sibling nodes and depth. Matsuzaki et al. proposed the PCFG with latent annotations (PCFG-LA) [Matsuzaki *et al.*05], which deal annotations as latent variables. In PCFG-LA, parameters of production rules are estimated by the expectation maximization (EM) algorithm. Since PCFG-LA is a more flexible model than PCFG with fixed annotations, we expect that PCFG-LA is more suitable for a base line grammar of GP-EDA. Therefore we have proposed GP-EDA based on PCFG-LA and named PAGE (programming with annotated grammar estimation).

5.4.6.2 PCFG with Latent Annotations

In this section, we briefly explain PCFG-LA, which is adopted as a base line grammar in our proposed method.

Matsuzaki et al. [Matsuzaki *et al.*05] proposed PCFG-LA, and they also derived a way of the parameter inference by means of EM algorithm. Although PCFG-LA adopted in PAGE has been modified for the present applications, it is essentially identical to the conventional PCFG-LA. Thus for further details of PCFG-LA, readers may refer to [Matsuzaki *et al.*05].

PCFG-LA assumes that every non-terminal has annotations. In the complete form, each non-terminal is denoted by $A[x]$, where A is non-terminal and x is an annotation attached to A. Fig. 5.25 describes an example of a tree with latent annotations (a) and their observed tree (b).

A likelihood of complete data (a tree with latent annotations) is represented by equation (5.32).

$$
\begin{aligned}
&P(T_i, X_i | \boldsymbol{\beta}, \boldsymbol{\pi}, h) \\
&= \prod_{x \in H} \pi(\mathcal{S}[x])^{\delta(x; T_i, X_i)} \\
&\quad \times \prod_{\mathcal{S}[x] \to \alpha \in \mathcal{R}[H]} \beta(\mathcal{S}[x] \to \alpha)^{c(\mathcal{S}[x] \to \alpha; T_i, X_i)}.
\end{aligned}
\tag{5.32}
$$

In this equation, H is a set of annotations, T_i is i-th derivation tree, X_i is a set of latent annotations of T_i ($X_i = \{x_i^1, x_i^1, \cdots\}$, where x_i^j is the j-th annotation of T_i), $\pi(\mathcal{S}[x])$ is the probability of $\mathcal{S}[x]$ at the root of T_i, $\beta(r)$ is the probability of a production rule $r \in \mathcal{R}[H]$, $\delta(x; T_i; X_i)$ is 1 if the annotation at the root position of T_i is x and 0 otherwise, $c(r; T_i, X_i)$ is the frequencies of rule r, h is the annotation size. $\boldsymbol{\beta} = \{\beta(\mathcal{S}[x] \to \alpha) | \mathcal{S}[x] \to \alpha \in \mathcal{R}[H]\}$ and $\boldsymbol{\pi} = \{\pi(\mathcal{S}[x]) | x \in H\}$. A set of annotated rules $\mathcal{R}[H]$ is given in equation (5.35).

A likelihood of an incomplete tree (an observed tree) can be calculated by summing over annotations:

$$
P(T_i | \boldsymbol{\beta}, \boldsymbol{\pi}, h) = \sum_{X_i} P(T_i, X_i | \boldsymbol{\beta}, \boldsymbol{\pi}, h).
\tag{5.33}
$$

The problem in PCFG-LA is to infer $\boldsymbol{\beta}$ and $\boldsymbol{\pi}$ (and also h for some cases).

5.4.6.3 Production Rules

In the proposed method PAGE, production rules are limited to a form represented by equation (5.34):

$$
\mathcal{S}[x] \to g\,\mathcal{S}[y]\,\mathcal{S}[y]\cdots\mathcal{S}[y],
\tag{5.34}
$$

where $x, y \in H$, and $g \in \mathcal{T}$. In these production rules, it is assumed that right-side non-terminals have the same annotation. The number of $\mathcal{S}$s in the right-side is arity of g. If $g = +$, then the production rule becomes $\mathcal{S}[x] \to +\,\mathcal{S}[y]\,\mathcal{S}[y]$. Let $\mathcal{R}[H]$ be a set of annotated rules expressed by equation (5.35):

$$
\mathcal{R}[H] = \{\mathcal{S}[x] \to g\,\mathcal{S}[y]\,\mathcal{S}[y]\cdots\mathcal{S}[y] | x, y, \in H, g \in \mathcal{T}\}.
\tag{5.35}
$$

We also define a set $\mathcal{R}_r[H]$ which is composed of the right-sides in $\mathcal{R}[H]$, i.e.,

$$
\mathcal{R}_r[H] = \{\alpha | \mathcal{S}[x] \to \alpha \in \mathcal{R}[H]\}.
\tag{5.36}
$$

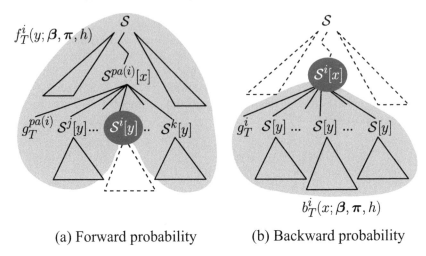

(a) Forward probability   (b) Backward probability

FIGURE 5.26: Forward f_T^i (a) and backward b_T^i (b) probabilities. The superscripts, which are attached to $\mathcal{S}$, are the indices of non-terminals (i in $\mathcal{S}^i[y]$ for example).

5.4.6.4   Forward-backward Probabilities

Since direct calculation of the EM algorithm is intractable because of computational cost, in [Matsuzaki *et al*.05], forward and backward probabilities are employed to perform the EM algorithm. Forward and backward probabilities are based on the dynamic programming. The backward probability $b_T^i(x; \boldsymbol{\beta}, \boldsymbol{\pi}, h)$ is a probability that a tree under i-th non-terminal is generated (Fig. 5.26 (b)), and is represented by equation (5.37):

$$b_T^i(x; \boldsymbol{\beta}, \boldsymbol{\pi}, h)$$
$$= \sum_{y \in H} \beta(\mathcal{S}[x] \to g_T^i \, \mathcal{S}[y] \cdots \mathcal{S}[y]) \prod_{j \in ch(i,T)} b_T^j(y; \boldsymbol{\beta}, \boldsymbol{\pi}, h).$$

$$(5.37)$$

In this equation, $ch(i, T)$ is a set of non-terminal child indices of the i-th non-terminal in T, $pa(i, T)$ is an index of a parent symbol of the i-th non-terminal in T, g_T^i is a terminal symbol in CFG and is directly connected to the i-th non-terminal symbol in T.

The forward probability $f_T^i(y; \boldsymbol{\beta}, \boldsymbol{\pi}, h)$ is a probability that a tree above i-th non-terminal is generated (Fig. 5.26(a)), and can be represented by equation (5.38):

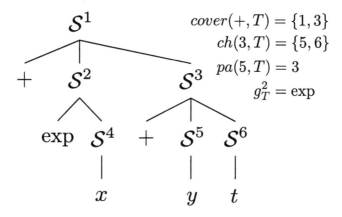

FIGURE 5.27: Examples of specific functions on a derivation tree.

$$
\begin{aligned}
&f_T^i(y; \boldsymbol{\beta}, \boldsymbol{\pi}, h) \\
&= \sum_{x \in H} f_T^{pa(i,T)}(x; \boldsymbol{\beta}, \boldsymbol{\pi}, h)\beta(\mathcal{S}[x] \to g_T^{pa(i,T)}\, \mathcal{S}[y] \cdots \mathcal{S}[y]) \\
&\quad \times \prod_{j \in ch(pa(i,T),T), j \neq i} b_T^j(y; \boldsymbol{\beta}, \boldsymbol{\pi}, h) \quad (i \neq 1),
\end{aligned}
\tag{5.38}
$$

$$
f_T^i(y; \boldsymbol{\beta}, \boldsymbol{\pi}, h) = \pi(\mathcal{S}[y]) \quad (i = 1).
\tag{5.39}
$$

With these two probabilities, a likelihood of an observed tree $P(T; \boldsymbol{\beta}, \boldsymbol{\pi}, h)$ can be represented by equations (5.40) and (5.41):

$$
P(T; \boldsymbol{\beta}, \boldsymbol{\pi}, h) = \sum_{x \in H} \pi(\mathcal{S}[x]) b_T^1(x; \boldsymbol{\beta}, \boldsymbol{\pi}, h),
\tag{5.40}
$$

$$
\begin{aligned}
&P(T; \boldsymbol{\beta}, \boldsymbol{\pi}, h) \\
&= \sum_{x,y \in H} \Big\{ \beta(\mathcal{S}[x] \to g\, \mathcal{S}[y] \cdots \mathcal{S}[y]) f_T^i(x; \boldsymbol{\beta}, \boldsymbol{\pi}, h) \\
&\quad \times \prod_{j \in ch(i,T)} b_T^j(y; \boldsymbol{\beta}, \boldsymbol{\pi}, h) \Big\} \quad (i \in cover(g, T)),
\end{aligned}
\tag{5.41}
$$

where $cover(g, T_i)$ is a set of non-terminal indices connected to g in T_i. We show an example of a concrete value of $cover(g, T_i)$ in Fig. 5.27.

Table 5.12: Comparison of two proposed methods: PAGE-EM and PAGE-VB.

Algorithm	Estimation	Target	Annotation size inference
PAGE-EM	EM algorithm	Parameter	Impossible
PAGE-VB	Variational Bayes	Distribution	Possible

5.4.6.5   Model Learning in PCFG-LA

Since PCFG-LA has latent variables, the model learning in PCFG-LA uses advanced statistical methods. [Matsuzaki *et al.*05] adopted EM algorithm for parameter inference. EM algorithm is a maximum likelihood estimator. If the annotation size h is known in advance, EM algorithm is well suited for model learning. However, it cannot be expected that the annotation size is available beforehand and it is desirable to infer not only the parameters β and π, but also the annotation size h. For inference of model parameters (in this case, this corresponds to the annotation size), Bayes estimation is used in general. The best model-parameter can be selected which yields the maximum posterior probability given learning data. For the case of models including latent variables, variational Bayes learning has been proposed [Attias99]. Variational Bayes learning can be considered to be a Bayesian EM algorithm. As a consequence, we also adopted variational Bayes for model learning of PCFG-LA.

In our method, we employed both EM algorithm and variational Bayes for model learning. To distinguish these two approaches, we labeled the former and the latter approaches by PAGE-EM and PAGE-VB, respectively. If we use a term "PAGE" in this chapter, then it refers to both PAGE-EM and PAGE-VG (Table 5.12).

5.4.6.5.1   PAGE-EM: EM Algorithm-based Estimation   We first describe the application of EM algorithm to PCFG-LA. Since the study, which originally proposed PCFG-LA adopted the EM algorithm [Matsuzaki *et al.*05], details of applications of EM algorithm for PCFG-LA are essentially identical to the original study.

Let $\overline{\beta}$ and $\overline{\pi}$ be updated parameters. EM algorithm estimates parameters which maximize the $\mathcal{Q}(\overline{\beta}, \overline{\pi}|\beta, \pi)$ function defined by equation (5.42).

$$\mathcal{Q}(\overline{\beta}, \overline{\pi}|\beta, \pi)$$
$$= \sum_{i=1}^{N} \sum_{X_i} P(X_i|T_i; \beta, \pi, h) \log P(T_i, X_i; \overline{\beta}, \overline{\pi}, h). \tag{5.42}$$

In order to maximize $\mathcal{Q}$ under the constraints that $\sum_{\alpha} \beta(\mathcal{S}[x] \to \alpha) = 1$ and $\sum_{x} \pi(\mathcal{S}(x)) = 1$, the Lagrange method is used. By applying the Lagrange

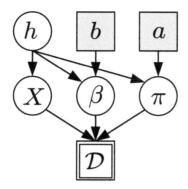

FIGURE 5.28: Graphical model for PCFG-LA with variational Bayes learning.

method, we finally obtain the following parameter update formulae:

$$\overline{\pi}(\mathcal{S}[x]) \propto \pi(\mathcal{S}[x]) \sum_{i=1}^{N} \frac{b_{T_i}^1(x; \boldsymbol{\beta}, \boldsymbol{\pi}, h)}{P(T_i; \boldsymbol{\beta}, \boldsymbol{\pi}, h)}, \tag{5.43}$$

$$\overline{\beta}(\mathcal{S}[x] \to g\,\mathcal{S}[y]\cdots\mathcal{S}[y]) \propto \beta(\mathcal{S}[x] \to g\,\mathcal{S}[y]\cdots\mathcal{S}[y])$$

$$\times \sum_{i=1}^{N} \frac{1}{P(T_i; \boldsymbol{\beta}, \boldsymbol{\pi}, h)} \sum_{j \in cover(g, T_i)} f_{T_i}^j(x; \boldsymbol{\beta}, \boldsymbol{\pi}, h)$$

$$\times \prod_{k \in ch(j, T_i)} b_{T_i}^k(y; \boldsymbol{\beta}, \boldsymbol{\pi}, h). \tag{5.44}$$

In these equations, N is the number of learning samples (in EDA, this corresponds to the number of promising solutions). EM algorithm monotonically increases a likelihood function by iteratively updating the parameters using equations (5.44) and (5.43).

5.4.6.5.2 PAGE-VB: Variational Bayes-based estimation

We next explain an application of variational Bayes to PCFG-LA. As mentioned before, by using variational Bayes learning, not only the parameters, but also the annotation size can be inferred. Fig. 5.28 depicts a graphical model of variational Bayes for PCFG-LA.

We assume that a prior distribution of $\boldsymbol{\beta}$ is a product of Dirichlet distribution (equations (5.45) and (5.46)). Dirichlet distribution is selected only due to computational reasons. When we use Dirichlet distribution for $\boldsymbol{\beta}$, the posterior of $\boldsymbol{\beta}$ can be calculated in closed form:

$$P(\boldsymbol{\beta}|h) = \prod_{x \in H} \mathrm{Dir}(\boldsymbol{\beta}_{\mathcal{S}[x]}; \mathbf{b}_{\mathcal{S}[x]}), \tag{5.45}$$

$$\mathrm{Dir}(\beta_{\mathcal{S}[x]}; \mathbf{b}_{\mathcal{S}[x]})$$

$$= \frac{\Gamma\left(\displaystyle\sum_{\alpha \in \mathcal{R}_r[H]} b_{\mathcal{S}[x] \to \alpha}\right)}{\displaystyle\prod_{\alpha \in \mathcal{R}_r[H]} \Gamma(b_{\mathcal{S}[x] \to \alpha})} \prod_{\alpha \in \mathcal{R}_r[H]} \beta(\mathcal{S}[x] \to \alpha)^{b_{\mathcal{S}[x] \to \alpha} - 1}.$$

$$(5.46)$$

In equation (5.46), $\Gamma(x)$ is the gamma function, $\mathrm{Dir}(x)$ is the Dirichlet distribution. We further defined variables, $\beta_{\mathcal{S}[x]} \equiv \{\beta(\mathcal{S}[x] \to \alpha)|\alpha \in \mathcal{R}_r[H]\}$ and $\beta \equiv \{\beta_{\mathcal{S}[x]}|x \in H\}$. $\mathbf{b}_{\mathcal{S}[x]}$ denotes the hyper-parameter of $\beta_{\mathcal{S}[x]}$.

We also assume that a prior of π is the Dirichlet distribution:

$$P(\pi|h) = \mathrm{Dir}(\pi; \mathbf{a}),$$

$$= \frac{\Gamma\left(\displaystyle\sum_{x \in H} a_{\mathcal{S}[x]}\right)}{\displaystyle\prod_{x \in H} \Gamma(a_{\mathcal{S}[x]})} \prod_{x \in H} \pi(\mathcal{S}[x])^{a_{\mathcal{S}[x]} - 1}. \qquad (5.47)$$

By calculating the posterior of β and π by following the procedures of variational Bayes, we obtain the posterior distributions. Posterior of β and π (which are expressed $Q(\cdot)^2$) can be represented by equations (5.48) and (5.49), respectively:

$$Q(\beta_{\mathcal{S}[x]}|h) \propto P(\beta_{\mathcal{S}[x]}|h) \times$$
$$\exp \langle \log P(\mathcal{D}, \mathbf{X}|\beta, \pi, h) \rangle_{Q(\mathbf{X}|h), Q(\beta_{-\mathcal{S}[x]}|h), Q(\pi|h)},$$

$$(5.48)$$

$$Q(\pi|h) \propto P(\pi|h) \exp \langle \log P(\mathcal{D}, \mathbf{X}|\beta, \pi, h) \rangle_{Q(\mathbf{X}|h), Q(\beta|h)}. \qquad (5.49)$$

In equations (5.48) and (5.49), $\langle \cdots \rangle_{Q(\cdots)}$ represent the expectation value with respect to $Q(\cdots)$, while $\langle \cdots \rangle_{Q(\beta_{-\mathcal{S}[x]}|h)}$ denotes the calculation of the expectation value with

---

[2] Since $Q(\cdot)$ is the (approximate) posterior distribution, it may be represented by $Q(\cdot|\mathcal{D})$. However, we use $Q(\cdot)$ for a notational convenience.

$$\langle \cdots \rangle_{Q(\boldsymbol{\beta}_{-\mathcal{S}[x]}|h)} \equiv$$

$$\int \left\{ \prod_{w \in H, w \neq x} d\boldsymbol{\beta}_{\mathcal{S}[w]} \right\} \left\{ \prod_{w \in H, w \neq x} Q(\boldsymbol{\beta}_{\mathcal{S}[w]}|h) \right\} (\cdots).$$

Calculating these equations, we finally obtain $Q(\boldsymbol{\beta}_{\mathcal{S}[x]}|h)$:

$$Q(\boldsymbol{\beta}_{\mathcal{S}[x]}|h) = \mathrm{Dir}(\boldsymbol{\beta}_{\mathcal{S}[x]}; \check{\mathbf{b}}_{\mathcal{S}[x]}), \tag{5.50}$$

$$\check{\mathbf{b}}_{\mathcal{S}[x]} \equiv \{\check{b}_{\mathcal{S}[x] \to \alpha} | \alpha \in \mathcal{R}_r[H]\}, \tag{5.51}$$

$$\check{b}_{\mathcal{S}[x] \to \alpha} = \sum_{i=1}^{N} \sum_{X_i} Q(X_i|h) \cdot c(\mathcal{S}[x] \to \alpha; T_i, X_i) + b_{\mathcal{S}[x] \to \alpha}, \tag{5.52}$$

and $Q(\boldsymbol{\pi}|h)$:

$$Q(\boldsymbol{\pi}|h) = \mathrm{Dir}(\boldsymbol{\pi}; \check{\mathbf{a}}), \tag{5.53}$$

$$\check{\mathbf{a}} \equiv \{\check{a}_{\mathcal{S}[x]} | x \in H\}, \tag{5.54}$$

$$\check{a}_{\mathcal{S}[x]} = \sum_{i=1}^{N} \sum_{X_i} Q(X_i|h) \cdot \delta(x; T_i, X_i) + a_{\mathcal{S}[x]}. \tag{5.55}$$

The posterior of latent variables can be calculated in a similar way:

$$Q(X_i|h) \propto \exp \langle \log P(T_i, X_i|\boldsymbol{\beta}, \boldsymbol{\pi}, h) \rangle_{Q(\boldsymbol{\pi}|h), Q(\boldsymbol{\beta}|h)}. \tag{5.56}$$

By calculating equation (5.56), a specific form of the posterior is represented below:

$$Q(X_i|h) =$$
$$\frac{1}{\mathcal{Z}(T_i)} \left\{ \prod_{x \in H} \widetilde{\pi}(\mathcal{S}[x])^{\delta(x; T_i, X_i)} \right\} \times$$
$$\left\{ \prod_{\mathcal{S}[x] \to \alpha \in \mathcal{R}[H]} \widetilde{\beta}(\mathcal{S}[x] \to \alpha)^{c(\mathcal{S}[x] \to \alpha; T_i, X_i)} \right\}, \tag{5.57}$$

$$\widetilde{\beta}(\mathcal{S}[x] \to \alpha) =\equiv \exp \left[\psi(\check{b}_{\mathcal{S}[x] \to \alpha}) - \psi \left(\sum_{\alpha \in \mathcal{R}_r[H]} \check{b}_{\mathcal{S}[x] \to \alpha} \right) \right],$$

$$\tilde{\pi}(\mathcal{S}[x]) \equiv \exp\left[\psi(\breve{a}_{\mathcal{S}[x]}) - \psi\left(\sum_{x \in H} \breve{a}_{\mathcal{S}[x]}\right)\right].$$

In these equations, $\psi(x)$[3] is the digamma function, and $\mathcal{Z}(T_i)$ is a normalizing constant. The normalizing constant is represented by equation (5.58).

$$\mathcal{Z}(T_i) = \sum_{x \in H} \tilde{\pi}(\mathcal{S}[x]) b^1_{T_i}(x; \tilde{\boldsymbol{\beta}}, \tilde{\boldsymbol{\pi}}, h). \tag{5.58}$$

In equations (5.52) and (5.55), it is intractable to directly calculate sum over latent annotations. These terms can be calculated using forward and backward probabilities:

$$\sum_{X_i} Q(X_i|h) \cdot c(\mathcal{S}[x] \to g\,\mathcal{S}[y]\,\mathcal{S}[y] \cdots \mathcal{S}[y]; T_i, X_i)$$

$$= \frac{\tilde{\beta}(\mathcal{S}[x] \to g\,\mathcal{S}[y]\,\mathcal{S}[y] \cdots \mathcal{S}[y])}{\mathcal{Z}(T_i)}$$

$$\times \sum_{k \in cover(g, T_i)} f^k_{T_i}(x; \tilde{\boldsymbol{\beta}}, \tilde{\boldsymbol{\pi}}, h) \prod_{j \in ch(k, T_i)} b^j_{T_i}(y; \tilde{\boldsymbol{\beta}}, \tilde{\boldsymbol{\pi}}, h), \tag{5.59}$$

$$\sum_{X_i} Q(X_i|h) \cdot \delta(x; T_i, X_i) = \frac{\tilde{\pi}(\mathcal{S}[x])}{\mathcal{Z}(T_i)} b^1_{T_i}(x; \tilde{\boldsymbol{\beta}}, \tilde{\boldsymbol{\pi}}, h). \tag{5.60}$$

We can see similarities between equations (5.59) - (5.60) , and equations (5.43) - (5.44).

5.4.6.6   Generation of New Individuals

As mentioned in Section 5.2.4, EDA generates new individuals according to the predictive posterior distribution (equation (5.5)). Let $\mathcal{P}_g$ be a population at generation g and $\mathcal{D}_g$ be a set of promising individuals (selected under a selection process) at generation g. In EDA, $\mathcal{D}_g$ is treated as learning data. PAGE generates a population of the next generation $\mathcal{P}_{g+1}$ by sampling from $P(T, X|\mathcal{D}_g)$. We separately explain the sampling process of PAGE-EM and PAGE-VB.

5.4.6.6.1   Sampling in PAGE-EM   The posterior probability for the case of EM algorithm can be represented by equation (5.61):

$$P(T, X|\mathcal{D}_g) = P(T, X|\overline{\boldsymbol{\beta}}^*, \overline{\boldsymbol{\pi}}^*, h). \tag{5.61}$$

---

[3] $\psi(x) \equiv \frac{\partial}{\partial x} \ln \Gamma(x)$

In this equation, $\overline{\beta}^*$ and $\overline{\pi}^*$ are converged parameters obtained by calculating parameters update formulae (equations (5.43) and (5.44)), and h is the annotation size given in advance. Since posterior in PAGE-EM is decomposed, PLS can be used to sample from the distribution.

5.4.6.6.2   Sampling in PAGE-VB   It is intractable to sample from the posterior of PAGE-VB because of computational reasons. Thus we adopted MAP approximation, and the posterior of PAGE-VB can be approximated by

$$P(T, X | \mathcal{D}_g) \simeq P(T, X | \beta^*_{\mathrm{MAP}}, \pi^*_{\mathrm{MAP}}, h^{opt}_g), \tag{5.62}$$

where h is the best annotation size. If the prior of the annotation size is $P(h) \propto 1$ (uniform), the best annotation size yields the maximum of $\mathcal{F}^*_h[Q]$,

$$Q^*(h) = \frac{P(h) \exp\{\mathcal{F}^*_h[Q]\}}{\displaystyle\sum_{\ell \in \mathcal{H}} P(\ell) \exp\{\mathcal{F}^*_\ell[Q]\}}. \tag{5.63}$$

Thus, the h^{opt}_g can be obtained by the following equation:

$$h^{opt}_g = \operatorname*{argmax}_h \mathcal{F}^*_h[Q].$$

β^*_{MAP} and π^*_{MAP} are MAP values of approximate posteriors $Q(\beta^* | h^{opt}_g)$ and $Q(\pi^* | h^{opt}_g)$, and are calculated by the Lagrange method.

5.4.6.7   Annotation Size Selection

In the previous section, we selected the optimal annotation size by calculating $\mathcal{F}^*_h[Q]$. This indicates that calculation of all $h \in \mathcal{H}$, $\mathcal{F}^*_1[Q], \mathcal{F}^*_2[Q], \mathcal{F}^*_3[Q], \cdots$ are required. However, selecting the optimum annotation size in this way is a highly computationally expensive method. As a consequence, we employed two heuristics listed below to reduce the computational costs.

- **Sparse Exploration**
  At the initial generation, we calculate $\mathcal{F}^*_h[Q]$ only sparsely. We used a variable $\mathcal{H}_0$ to be a set of annotation sizes to explore for the initial generation ($g = 0$).

- **Exploration Restriction**
  It is natural to think that the $\mathcal{D}_g$ of neighboring generations are very similar. We take advantage of this assumption. At generation $g + 1$, PAGE-VB only calculates the h_{range} neighboring annotation sizes of h^{opt}_g obtained for generation g.

Fig. 5.29 shows the heuristics. In this figure, g is a generation. Gray cells represent h for which $\mathcal{F}^*_h[Q]$ is calculated. Black cells are the optimum annotation size.

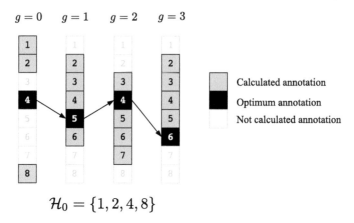

$$\mathcal{H}_0 = \{1, 2, 4, 8\}$$

FIGURE 5.29: An intuitive figure explaining two heuristics adopted in PAGE-VB. g represents the generation, and gray cells are those whose corresponding values of $\mathcal{F}_h^*[Q]$ are calculated, while black cells denote the optimum annotation size h_g^{opt}.

5.4.6.8 Experiments

We applied PAGE to several benchmark tests to investigate the behavior and to show the effectiveness of our approach. We used the royal tree problem (level d), the royal tree problem (level e), the DMAX problem (weak deceptiveness version), and the DMAX problem (strong deceptiveness version). At each benchmark test, we performed 50 runs. Tables 5.13 and 5.14 show parameter settings for PAGE-EM and PAGE-VB, respectively.

5.4.6.8.1 Royal Tree Problem (level d)   We first applied our approach to the royal tree problem so as to investigate the effect of annotation size on search performance. We used three approaches listed below in this experiment.

- **PAGE-EM**
  In this experiment, we considered four values $h = 2, 4, 8, 16$.

- **PAGE-VB**
  We set the possible annotation sizes for the initial generation to be $\mathcal{H}_0 = \{1, 2, 4, 8, 16\}$.

- **PCFG-GP**
  In this section, PCFG-GP denotes a GP-EDA based on the conventional PCFG.

For the present test problem, we employed a level d royal tree problem. For details of the royal tree problem, see Section 5.2.6. Production rules for this test problem are represented below. In these rules, a, b, c, d are function nodes

Table 5.13: Main parameters for PAGE-EM.

Variable	Parameter	Value
M	Population size	1,000 (Royal Tree level d) 2,500 (Royal Tree level e) 1,000 (DMAX weak) 1,000 (DMAX strong)
h	Annotation size	2,4,8,16 (Royal Tree level d) 16 (Royal Tree level e) 16 (DMAX weak) 8 (DMAX strong)
D_P	Maximum depth	5 (Royal Tree level d) 6 (Royal Tree level e) 5 (DMAX weak) 4 (DMAX strong)
P_s	Selection rate	0.1 (Royal Tree level d) 0.05 (Royal Tree level e) 0.1 (DMAX weak) 0.1 (DMAX strong)
P_e	Elite rate	0.1 (Royal Tree level d) 0.05 (Royal Tree level e) 0.1 (DMAX weak) 0.1 (DMAX strong)

and are identical to A, B, C, D in Section 5.2.6. We used lower cases to emphasize that function nodes in GP are terminal symbols in PCFG.

$$\mathcal{S} \rightarrow a\,\mathcal{S},$$
$$\mathcal{S} \rightarrow b\,\mathcal{S}\,\mathcal{S},$$
$$\mathcal{S} \rightarrow c\,\mathcal{S}\,\mathcal{S}\,\mathcal{S},$$
$$\mathcal{S} \rightarrow d\,\mathcal{S}\,\mathcal{S}\,\mathcal{S}\,\mathcal{S},$$
$$\mathcal{S} \rightarrow x.$$

Results Fig. 5.30 shows the cumulative frequency of successful runs at each fitness evaluation. PCFG-GP which does not use latent annotations failed to obtain the optimum of the royal tree problem (level d). Since PCFG-GP simply counts the frequencies of each production rule, the rule $\mathcal{S} \rightarrow x$ has the highest probability. Using this probability distribution for generating new individuals, only small trees are generated because probabilities of large trees tend to be very small. Because PAGE-EM can consider the dependencies among nodes using latent annotations, large derivation trees can be generated. Table 5.15 describes production rules inferred by PAGE-EM. We show the production rules of $h = 8$ and $h = 16$. As can been seen with this table,

Table 5.14: Main parameters for PAGE-VB.

Variable	Parameter	Value
M	Population size	1,000 (Royal Tree level d)
		4,000 (Royal Tree level e)
		1,000 (DMAX weak)
		1,000 (DMAX strong)
$\mathcal{H}_0$	Annotation to explore	$\{1, 2, 4, 8, 16\}$
h_{rage}	Annotation exploration range	2
D_P	Maximum depth	5 (Royal Tree level d)
		6 (Royal Tree level e)
		5 (DMAX weak)
		4 (DMAX strong)
P_s	Selection rate	0.1 (Royal Tree level d)
		0.05 (Royal Tree level e)
		0.1 (DMAX weak)
		0.1 (DMAX strong)
P_e	Elite rate	0.1 (Royal Tree level d)
		0.05 (Royal Tree level e)
		0.1 (DMAX weak)
		0.1 (DMAX strong)

production rules are specialized by using annotations. For instance in $h = 16$, $\mathcal{S}[0]$, $\mathcal{S}[1]$, $\mathcal{S}[2]$, $\mathcal{S}[9]$, $\mathcal{S}[12]$, $\mathcal{S}[13]$, and $\mathcal{S}[14]$ generate x exclusively and $\mathcal{S}[5]$ generates d exclusively. Because of this specialization, PAGE can infer the building blocks.

We can see from Fig. 5.30 that the number of annotations strongly affects the search performance in PAGE-EM. As can been seen, larger annotation size yielded the better performance. The production rules with larger annotation size are more specialized and it can estimate the building blocks more correctly.

The search performance of PAGE-VB is worse than that of PAGE-EM. However, this fact does not mean that PAGE-VB is inferior to PAGE-EM. Since PAGE-EM has to decide the annotation size in advance, it requires extra computational costs. That is, trial and error are required to determine the optimum annotations. On the other hand, PAGE-VB automatically estimates the annotation size by calculating $\mathcal{F}_h^*[Q]$.

Fig. 5.31 describes $\mathcal{F}_h^*[Q]$ for each annotation size at generations 0 and 17. As can been seen, $\mathcal{F}_h^*[Q]$ is peaked at $h = 5$ (generation 0) and $h = 6$ (generation 17). PAGE-VB selected these annotation sizes as the optimum annotation sizes for generations 0 and 17. With this experiment, we showed that larger annotation size yielded better performance. On the other hand, larger annotation size required more computational power. PAGE-VB balanced the performance against the computational costs.

Fig. 5.32 shows the transition of log-likelihood in the royal tree problem

Table 5.15: Parameters estimated for the royal tree problem (level d) by PAGE-EM with $h = 8$ and $h = 16$.

PAGE-EM ($h = 16$)	Probability
$S[5]$	1.000
$S[0] \to x$	1.000
$S[1] \to x$	1.000
$S[2] \to x$	1.000
$S[3] \to c\,S[13]\,S[13]\,S[13]$	0.063
$S[3] \to c\,S[14]\,S[14]\,S[14]$	0.131
$S[3] \to c\,S[9]\,S[9]\,S[9]$	0.093
$S[3] \to$	0.054
$d\,S[14]\,S[14]\,S[14]\,S[14]$	
$S[3] \to d\,S[9]\,S[9]\,S[9]\,S[9]$	0.060
$S[3] \to x$	0.329
$S[4] \to a\,S[1]$	0.154
$S[4] \to a\,S[14]$	0.257
$S[4] \to a\,S[2]$	0.189
$S[4] \to a\,S[9]$	0.238
$S[5] \to$	1.000
$d\,S[10]\,S[10]\,S[10]\,S[10]$	
$S[6] \to a\,S[14]$	0.144
$S[6] \to a\,S[9]$	0.092
$S[6] \to b\,S[1]\,S[1]$	0.070
$S[6] \to b\,S[2]\,S[2]$	0.079
$S[6] \to$	0.167
$d\,S[14]\,S[14]\,S[14]\,S[14]$	
$S[7] \to a\,S[14]$	0.152
$S[7] \to a\,S[9]$	0.184
$S[7] \to c\,S[9]\,S[9]\,S[9]$	0.071
$S[7] \to x$	0.319
$S[8] \to b\,S[11]\,S[11]$	0.188
$S[8] \to b\,S[4]\,S[4]$	0.397
$S[8] \to b\,S[6]\,S[6]$	0.079
$S[8] \to c\,S[3]\,S[3]\,S[3]$	0.053
$S[9] \to x$	1.000
$S[10] \to c\,S[15]\,S[15]\,S[15]$	0.900
$S[10] \to c\,S[8]\,S[8]\,S[8]$	0.093
$S[11] \to a\,S[1]$	0.167
$S[11] \to a\,S[12]$	0.080
$S[11] \to a\,S[13]$	0.181
$S[11] \to a\,S[14]$	0.209
$S[11] \to a\,S[2]$	0.143
$S[11] \to a\,S[9]$	0.208
$S[12] \to x$	0.993
$S[13] \to x$	1.000
$S[14] \to x$	1.000
$S[15] \to b\,S[11]\,S[11]$	0.648
$S[15] \to b\,S[4]\,S[4]$	0.239

PAGE-EM ($h = 8$)	Probability
$S[3]$	1.000
$S[0] \to b\,S[6]\,S[6]$	0.018
$S[0] \to c\,S[6]\,S[6]\,S[6]$	0.972
$S[1] \to x$	1.000
$S[2] \to a\,S[1]$	0.090
$S[2] \to a\,S[4]$	0.538
$S[2] \to a\,S[7]$	0.075
$S[2] \to b\,S[1]\,S[1]$	0.014
$S[2] \to b\,S[4]\,S[4]$	0.022
$S[2] \to b\,S[7]\,S[7]$	0.025
$S[2] \to c\,S[4]\,S[4]\,S[4]$	0.059
$S[2] \to c\,S[7]\,S[7]\,S[7]$	0.032
$S[2] \to d\,S[4]\,S[4]\,S[4]\,S[4]$	0.068
$S[2] \to x$	0.061
$S[3] \to d\,S[0]\,S[0]\,S[0]\,S[0]$	1.000
$S[4] \to x$	1.000
$S[5] \to a\,S[1]$	0.161
$S[5] \to a\,S[4]$	0.468
$S[5] \to a\,S[7]$	0.356
$S[6] \to a\,S[2]$	0.012
$S[6] \to b\,S[2]\,S[2]$	0.467
$S[6] \to b\,S[5]\,S[5]$	0.399
$S[6] \to c\,S[2]\,S[2]\,S[2]$	0.030
$S[6] \to d\,S[2]\,S[2]\,S[2]\,S[2]$	0.055
$S[6] \to x$	0.029
$S[7] \to x$	1.000

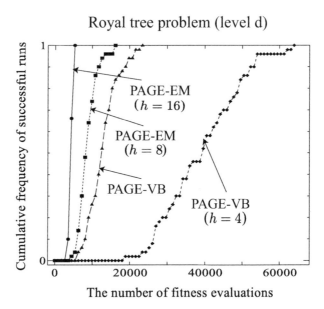

FIGURE 5.30: The number of fitness evaluations vs. cumulative frequency of successful runs for the royal tree problem at level d.

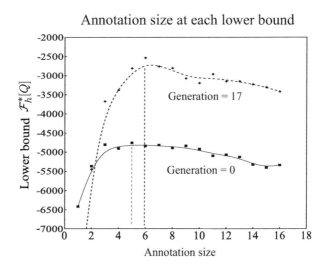

FIGURE 5.31: The converged values of the lower bound $\mathcal{F}_h^*[Q]$ against annotation size for the royal tree problem at level d.

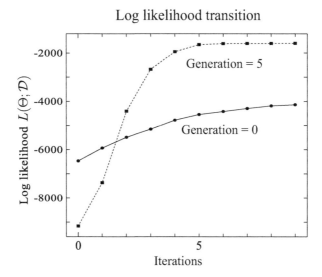

FIGURE 5.32: Transition of log-likelihood for the royal tree problem (level *d*) solved by PAGE-EM ($h = 16$). We show the transition for generations 0 and 5 (end of the search).

at level *d* using PAGE-EM. We can see that EM algorithm monotonically increased the log likelihood. We can see that parameters converged after around 10 iterations. The improvement of the log-likelihood at generation 5 is larger than that at generation 0. Since learning data converges, log-likelihood tends to be larger in the later generations.

Variational Bayes also maximizes the lower bound $\mathcal{F}_h^*[Q]$. Fig. 5.33 shows the transition of the lower bound at generation 17 and generation 0. We can see that variational Bayes also increased the lower bound monotonically.

5.4.6.8.2   Royal Tree Problem (level *e*)   We next applied PAGE to the royal tree problem for level *e*. We compared PAGE with GMPE (see Section 5.4.4). We used the following four algorithms for comparison.

- **PAGE-EM**
  In this experiment, we set $h = 16$ and $M = 2500$.

- **PAGE-VB**
  We set $\mathcal{H}_0 = \{1, 2, 4, 8, 16\}$ and $M = 4000$.

- **GMPE**
  For details of GMPE, please see Section 5.4.4. The result of GMPE is scanned from Fig. 2 in [Shan *et al*.04]. In [Shan *et al*.04], values of $M = 60$ (population size), and $P_s = 0.5$ (selection rate, a truncation selection) were used.

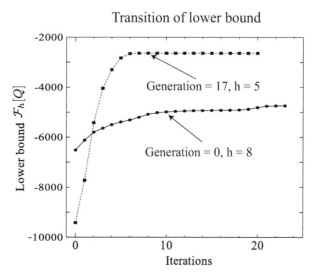

FIGURE 5.33: Change of the lower bound $\mathcal{F}_h[Q]$ at each iteration for the royal tree problem (level d).

In this experiment, we used the royal tree problem of level e. The level e problem is much more difficult than the level d. The number of nodes for the level e is 326, and five times larger than that of the level d. Production rules for the royal tree problem at level e are represented below.

$$\mathcal{S} \rightarrow a\,\mathcal{S},$$
$$\mathcal{S} \rightarrow b\,\mathcal{S}\,\mathcal{S},$$
$$\mathcal{S} \rightarrow c\,\mathcal{S}\,\mathcal{S}\,\mathcal{S},$$
$$\mathcal{S} \rightarrow d\,\mathcal{S}\,\mathcal{S}\,\mathcal{S}\,\mathcal{S},$$
$$\mathcal{S} \rightarrow e\,\mathcal{S}\,\mathcal{S}\,\mathcal{S}\,\mathcal{S}\,\mathcal{S},$$
$$\mathcal{S} \rightarrow x.$$

Results Table 5.16 shows the number of fitness evaluations required to obtain the optimum of the royal tree problem at level e. Since PAGE-VB and GMPE did not obtain the optimum at the probability of 100%, the top 90% of the runs are used for calculating the average. Fig. 5.34 plots the cumulative frequency of successful runs at each fitness evaluation. From this figure, PAGE-EM ($h = 16$) shows the best performance. Although half the runs of GMPE obtained the optimum very effectively, the rest required more fitness evaluations. With this result, it is considered that GMPE is affected by the initial population. Although GMPE used the population size of $M = 60$, PAGE-EM and PAGE-VB used $M = 2500$ and 4000, respectively. The use of

Table 5.16: The number of average fitness evaluations per one optimum for the royal tree problem at level e (these values are calculated using the top 90% of results).

Algorithm	The number of fitness evaluations	Standard deviation
PAGE-EM	61,809	15,341
PAGE-VB	263,720	63,528
GMPE	90,525	79,774

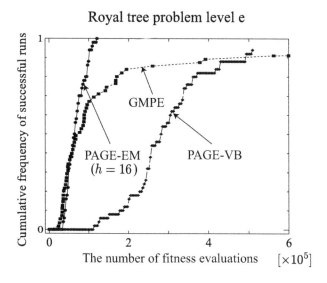

FIGURE 5.34: The number of fitness evaluations vs. the cumulative frequency of successful runs for the royal tree problem at level e.

more individuals makes the bias of individuals smaller, and as a result, PAGE was less vulnerable to the initial population. As depicted in Section 5.4.4, GMPE uses a very flexible model, which estimates not only the parameters but also production rules themselves from promising solutions. It can be considered that because the production rule inference is highly computationally expensive, GMPE adopted the smaller population size compared to PAGE. On the other hand, PAGE takes advantage of the approximation that the right-side non-terminals have the same annotation (equation (5.34)). Because of this approximation, the computational cost of PAGE is proportional to h^2 and PAGE can handle more individuals. Table 5.17 is the results of the statistical test. Each value represents p-values.

5.4.6.8.3 The DMAX problem (Weak Deceptiveness Version) We applied PAGE to the DMAX problem (Section 5.2.6), which is known as a GP-

Table 5.17: t-test for the royal tree problem at level e. Each value represents a p-value.

	PAGE-EM	PAGE-VB
PAGE-EM		5.90×10^{-23}
PAGE-VB	$-$	

hard benchmark test. In this experiment, we compare the search performance of PAGE with GP and POLE (see Section 5.3.4).

- **PAGE-EM**
  In this experiment, we set $h = 16$ and $M = 1000$.

- **PAGE-VB**
  We set $\mathcal{H}_0 = \{1, 2, 4, 8, 16\}$ and $M = 1000$.

- **PCFG-GP**
  The population size was again set to $M = 1000$.

- **Simple GP (GP)**
  In the experiments, $P_e = 0.01$ (elite rate), $P_c = 0.9$ (crossover rate), $P_m = 0$ (mutation rate), $P_r = 0.09$ (reproduction rate), the tournament size $t_s = 2$ and population size $M = 1000$ were used.

- **POLE**
  We set $M = 4000$, $R_P = 2$ (parent range), $P_e = 0.005$, $P_s = 0.1$ and a truncation selection was used.

In this section, we first consider a case with $m = 3$, $r = 2$ and $D_P = 5$, which has relatively weak deceptiveness. In this section, we denote this setting as the DMAX problem with weak deceptiveness. The production rules used in this experiment are represented below:

$$S \to +_3 \mathcal{S} \mathcal{S} \mathcal{S},$$
$$S \to \times_3 \mathcal{S} \mathcal{S} \mathcal{S},$$
$$S \to 0.95,$$
$$S \to -1.$$

Results Table 5.18 shows the average number of fitness evaluations for the various methods. Fig. 5.35 plots the cumulative frequency of successful runs at each fitness evaluation. Clearly, PAGE-EM and PAGE-VB outperformed GP and POLE. As mentioned in Section 5.2.6, the DMAX problem is deceptive when using the crossover operator. Because of this reason, GP required many fitness evaluations to obtain the optimum. In Section 5.3.4.8, we have shown

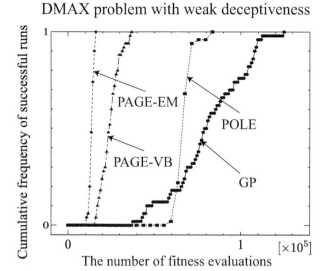

FIGURE 5.35: The number of fitness evaluations vs. cumulative frequency for successful runs for the DMAX problem with weak deceptiveness.

Table 5.18: The number of average fitness evaluations for one optimum in the case of DMAX problem with weak deceptiveness.

Algorithm	The number of fitness evaluations	Standard deviation
PAGE-EM	14,922	1,287
PAGE-VB	25,542	5,229
PCFG-GP	−	−
GP	82,332	20,874
POLE	72,516	4,381

the search performance of POLE in the DMAX problem and have argued that POLE is more effective than GP. Although POLE required less fitness evaluations compared to GP, it was inferior to PAGE-EM and PAGE-VB. This is because the prototype tree-based GP-EDA cannot infer the building blocks which are independent of their positions. On the other hand, PAGE can estimate the position-independent building blocks. Thus PAGE can reuse the estimated building blocks, and this difference explains the superiority of PAGE against POLE in this experiment.

Table 5.19 shows the results of statistical test. In this table, each value denotes p-value for the differences in means between PAGE-EM, PAGE-VB and GP, POLE.

Table 5.19: t-test for DMAX problem with weak deceptiveness. Each value represents a p-value.

	PAGE-EM	PAGE-VB	GP	POLE
PAGE-EM		1.68×10^{-19}	1.16×10^{-27}	5.28×10^{-63}
PAGE-VB	–		2.79×10^{-25}	1.27×10^{-68}
GP	–	–		2.17×10^{-3}
POLE	–	–	–	

5.4.6.8.4 DMAX problem (Strong Deceptiveness Version) We next applied PAGE to the DMAX problem with strong deceptiveness. Specifically, we compare the following methods:

- **PAGE-EM**
  We set $h = 8$ and $M = 1000$.

- **PAGE-VB**
  We set $\mathcal{H}_0 = \{1, 2, 4, 8, 16\}$ and $M = 1000$.

- **Simple GP (GP)**
  The population size is set to $M = 16000$. Other parameters are $P_e = 0.005$, $P_c = 0.995$, $P_m = 0$, $P_r = 0$, and $t_s = 2$.

- **POLE**
  The population size was set to $M = 6300$.

As mentioned in Section 5.2.6, the difficulty of the DMAX problem can be tuned with three parameters. In this section, we experimented with $m = 5, r = 3, D_P = 4$ with which this problem exhibits strong deceptiveness. The production rules used in this setting are listed below:

$$S \rightarrow +_5 \mathcal{S} \mathcal{S} \mathcal{S} \mathcal{S},$$
$$S \rightarrow \times_5 \mathcal{S} \mathcal{S} \mathcal{S} \mathcal{S},$$
$$S \rightarrow 0.95,$$
$$S \rightarrow \lambda.$$

Results Table 5.20 shows the average number of fitness evaluations. Since some algorithms did not obtain the optimum at the probability of 100%, we only used the top 90% of results to calculate averages. In this problem, two types of building blocks have to be used: $(0.95 + 0.95 + 0.95 + 0.95 + 0.95)$ and $(\lambda + \lambda + \lambda + \lambda + \lambda)$. Since PAGE uses only one non-terminal $\mathcal{S}$, production rules which generate 0.95 and λ have the same left non-terminal not considering annotations. Since PAGE adopts the latent annotations, 0.95 and λ are

Table 5.20: The number of average fitness evaluations per one optimum of DMAX problem with strong deceptiveness (these values are calculated using the top 90% of the runs).

	The number of fitness evaluations	Standard deviation
PAGE-EM	16,043	1,189
PAGE-VB	18,293	1,688
PCFG-GP	–	–
POLE	119,572	4,295
GP	1,477,376	293,890

generated from different annotations. Table 5.21 details the production rules inferred by PAGE-EM ($h = 8$). From this table, different annotations are used to generate 0.95 and λ. For example, non-terminals $S[0], S[1], S[3], S[5]$ generate λ, on the other hand $S[7]$ generates 0.95. By using these production rules, two building blocks are generated with efficiency.

Table 5.22 details the results of the t-tests. According to Table 5.22, the p-values for differences between PAGE-EM, PAGE-VB, GP and POLE means are extremely small, indicating that the differences between these means are statistically significant.

Fig. 5.36 shows the relation between the cumulative frequency of successful runs and the number of fitness evaluations. PAGE-EM and PAGE-VB showed a much better performance compared to GP and POLE. We have compared a search performance between PAGE and GP in the DMAX problem with weak deceptiveness. As can been seen, the performance difference in the DMAX problem with strong deceptiveness is larger. As shown in Section 5.2.6, the DMAX problem used in this section has very strong deceptiveness when using a crossover operator in GP. Because PAGE effectively estimates the building blocks, the deceptiveness of DMAX problem with strong deceptiveness can be escaped.

Similar to the case of DMAX with weak deceptiveness, POLE is inferior to PAGE. The prototype-tree approaches in general cannot estimate building blocks which are independent of position and this is the reason why POLE performed worse than PAGE.

5.4.6.9 Discussion

Because PCFG-LA uses a latent variable model, EM algorithm and variational Bayes are required for the model learning. For the case of EM algorithm, annotation size has to be given in advance. From the experiment of Section 5.4.6.8, it has been shown that higher annotation size yielded a better performance. However, higher annotation size requires more computational costs and is also faced with an over-fitting problem. In general, annotation size is not known in advance. Therefore trial and error are required to de-

Table 5.21: Estimated parameters for the DMAX problem with strong deceptiveness by PAGE-EM ($h = 8$). Rules with probabilities smaller than 0.01 are not shown because of limitations of space.

PAGE-EM ($h = 8$)	Probability
$S[4]$	1.000
$S[0] \rightarrow \lambda$	0.983
$S[0] \rightarrow 0.95$	0.017
$S[1] \rightarrow \lambda$	0.941
$S[1] \rightarrow 0.95$	0.059
$S[2] \rightarrow \times_5 S[6]\, S[6]\, S[6]\, S[6]\, S[6]$	1.000
$S[3] \rightarrow \lambda$	0.989
$S[3] \rightarrow 0.95$	0.011
$S[4] \rightarrow \times_5 S[2]\, S[2]\, S[2]\, S[2]\, S[2]$	1.000
$S[5] \rightarrow \lambda$	0.995
$S[6] \rightarrow +_5 S[0]\, S[0]\, S[0]\, S[0]\, S[0]$	0.399
$S[6] \rightarrow +_5 S[1]\, S[1]\, S[1]\, S[1]\, S[1]$	0.010
$S[6] \rightarrow +_5 S[3]\, S[3]\, S[3]\, S[3]\, S[3]$	0.234
$S[6] \rightarrow +_5 S[5]\, S[5]\, S[5]\, S[5]\, S[5]$	0.096
$S[6] \rightarrow +_5 S[7]\, S[7]\, S[7]\, S[7]\, S[7]$	0.261
$S[7] \rightarrow \lambda$	0.025
$S[7] \rightarrow 0.95$	0.975

cide the optimum annotation size in PAGE-EM. PAGE-VB uses variational Bayes for a model inference. Since variational Bayes inherits the properties of Bayesian estimation, variational Bayes is capable of selecting the best model. Fig. 5.37 shows the average number of estimated optimum annotation size for each problem. We divided generations in each run into three intervals: an early phase, a middle phase, and the end of phase. We can see from this figure that PAGE-VB estimated different annotation size in each problem. In the royal tree problem at level e, PAGE-VB estimated the largest annotation size. Since the population size in this problem is larger than those of other problems, PAGE-VB can use more learning data for model inference.

Table 5.22: t-test for the DMAX problem with strong deceptiveness. Each value represents a p-value.

	PAGE-EM	PAGE-VB	GP	POLE
PAGE-EM		2.82×10^{-9}	4.62×10^{-29}	1.10×10^{-61}
PAGE-VB	–		4.89×10^{-29}	6.58×10^{-67}
GP	–	–		7.13×10^{-28}
POLE	–	–	–	

DMAX problem with strong deceptiveness

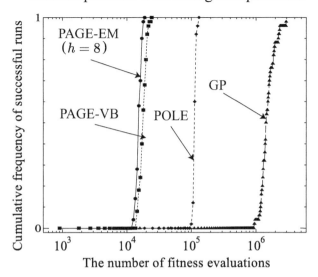

FIGURE 5.36: The number of fitness evaluations vs. cumulative frequency of successful runs for the DMAX problem (strong deceptiveness): the horizontal axis is in a log scale. Results for PCFG-GP are not presented since this method could not obtain the optimum.

Generally, more complex models can be estimated if more learning data are given. Furthermore, the royal tree problem at level e has more function nodes. For grasping building blocks with many function nodes, more annotations are required. Variational Bayes estimation also inferred the problem complexity of each level from learning data. For every problem, PAGE-VB estimated larger annotation size for later generations. In the later generations, since the promising solutions converge to the optimum structures, more specialized production rules can be used. We can say that PAGE-VB is a more effective approach than PAGE-EM considering that it is difficult and computationally expensive to decide the optimum annotation size.

5.4.7 Summary of PCFG-based GP-EDAs

We have proposed a probabilistic program evolution algorithm termed PAGE. Our proposal takes advantage of latent annotations that allow the context freedom assumption for a CFG to be weakened. By applying the proposed method to four computational experiments, we have confirmed that latent annotations are highly effective for determining the building blocks. For the royal tree problem (level d), we have showed that the number of annotations greatly affects the search performance, with larger annotation sizes offering

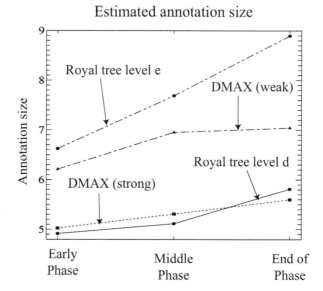

FIGURE 5.37: The optimum annotation size inferred by PAGE-VB for each problem.

better performance. The results for the DMAX problem have shown that PAGE is also highly effective for problems with strong deceptiveness. We hope that it will be possible to apply POLE to a wide class of real-world problems, which is an intended future area of study.

5.5   Other Related Methods

There are other GP-EDAs not belonging to either prototype tree or PCFG-based approaches. We specifically make a brief explanation for N-gram GP.

5.5.1 N-gram GP

Recently, a new type of GP-EDA has been proposed, and is called N-gram GP [Poli and McPhee08]. The term "N-gram" was first used in the field of NLP. N-gram considers subsequences in a given sequence. N-gram has customized names for the case of $N = 1, 2, 3$, and is called "uni-gram", "bi-gram", and "tri-gram", respectively. N-gram considers conditional probabilities of a given sequence. Let s_i be the i-th word in the sequence. N-gram estimates the conditional probabilities represented by equation (5.64):

$$P(s_j|s_{j-1}, s_{j-2}, \cdots, s_{j-N+1}). \tag{5.64}$$

Since N-gram is not accompanied with structure learning which is very computationally expensive, N-gram has been applied to various fields, e.g, bioinformatics, search engines and so on. N-gram GP takes advantage of this concept and applies N-gram to the linear GP. The linear GP is proposed by Nordin and does not use tree representations [Nordin94].

5.6 Summary

In this chapter, we introduced several GP-EDAs. Especially, we made a detailed explanation for POLE and PAGE. GP has been applied to a variety of problems because of its flexibility. On the other hand, GP based on statistical techniques has not been well developed until now. As demands for artificial intelligence grow, we think methods for automatic programming become more important. GA-EDA has been applied to real-world applications, including GA-hard problems. We expect that GP-EDA will be powerful methods for real-world problems which traditional GP cannot solve. In order to apply GP-EDA to real-world problems, we have to overcome many problems. Especially, we think that the following problems are the most important among them.

- Effect of introns

- Convergence and related theoretical analysis

- Selection of the best statistical model

- Killer applications of GP-EDA

Introns are structures which do not affect fitness values but are considered during model learning. If we apply model learning to individuals including many introns, it is difficult to infer the building blocks correctly. Many problems, to which GP is applied, contain introns. These problems include the symbolic regression and the parity problem. In order to make GP-EDA work well in these problems, GP-EDA has to treat introns. The simplest idea is to remove introns from tree structures. For example, techniques introduced by [Shin *et al.*07] can be considered. These methods remove introns using removal rules which are defined with heuristics.

As mentioned in this chapter, all EDAs assume statistical models in advance (Univariate, Bayesian Network, etc). However, it is very difficult to select the best statistical model before knowing the detailed problem to solve. Thus an automatic model selection in EDA is a very important problem.

Recently, Tanji and Iba have proposed a new GP-EDA method, which is motivated by the idea of preservation and control of tree fragments. They hypothesized that to reconstruct building blocks efficiently, tree fragments of any size should be preserved into the next generation, according to their differential fitnesses. This method creates a new individual by sampling from the promising trees by traversing and transition between trees instead of subtree crossover and mutation. Because the size of a fragment preserved during a generation update follows a geometric distribution, merits of the method are that it is relatively easy to predict the behavior of tree fragments over time and to control sampling size, by changing a single parameter. They have reported that there is a significant difference of fragment distribution between their method and the traditional GPs [Tanji and Iba09].

It is very important to show the effectiveness of GP-EDAs from the viewpoint of the number of fitness evaluations, to show that GP-EDAs require fewer fitness evaluations compared to that of GP. However, it is more appealing if GP-EDAs can obtain informations of target problems which traditional GPs cannot. For example, inferred graphical models (e.g, Bayesian network) or estimated parameters are knowledges which only GP-EDAs can obtain. Thus if there exists problems where estimated networks or estimated parameters themselves are important, GP-EDAs are possibly the killer algorithms for such problems. In such problems, it can be considered that GP-EDA may not be only used as an optimization algorithm, but also as a data mining tool. In recent years, GP has been applied to problems with huge amounts of data, like bioinformatics. If we can use inferred networks or estimated parameters as additional information, we hope that it is easier to analyze these data.

Appendix A

GUI Systems and Source Codes

A.1 Introduction

For the sake of better understanding GP and its extensions, software described in this book can be downloaded from the website of the laboratory to which one of the authors belongs (http://www.iba.t.u-tokyo.ac.jp/). They are LGPC, STROGANOFF v0.1, MVGPC, and EDA.

The intention of making these software packages available is to allow users to "play" with GA and GP. We request interested readers to download these packages and experience GP searches for themselves. The web pages contain (1) instructions for downloading the packages, (2) a user's manual, and (3) a bulletin board for posting questions. Readers are welcome to use these facilities.

Users who download and use these programs should bear the following items in mind:

1. We accept no responsibility for any damage that may be caused by downloading these programs.

2. We accept no responsibility for any damage that may be caused by running these programs or by the results of executing such programs.

3. We own the copyright to these programs.

4. We reserve the right to alter, add to, or remove these programs without notice.

In order to download the software in this book, please follow the next link, which contains further instructions:

http://www.iba.t.u-tokyo.ac.jp/english/BeforeDownload.htm

If inappropriate values are used for the parameters (for instance, elite size greater than population size), the program may crash. Please report any bugs found to: stroganoff@iba.t.u-tokyo.ac.jp.

A.2   STROGANOFF for Time Series Prediction

There are two versions of STROGANOFF v0.1 included. STROGANOFF for Time Series Prediction and STROGANOFF for System Identification. The two versions of the program are very similar, with the main difference being the format of the input data and the behavior of a few parameters.

STROGANOFF (Structured Representation On Genetic Algorithm for Nonlinear Function Fitting) was developed based on the work described by Iba and Nikolaev on Inductive Genetic Programming (see Chapter 3 for details). This software implements the technique, and a Graphical Interface to use it with any sort of Time Series or System Identification data.

STROGANOFF also allows you to run up to four populations in parallel, with different parameters, in order to see what is the comparative influence of those parameters in the evolutionary process.

A.2.1   System Requirements

Below are the system specifications used during the development of STRO-GANOFF. These may be considered minimum system requirements for the program. However, larger populations, more populations or more generations may require a larger amount of system resources.

OS: Windows XP, or WINE Emulator for Unix-like systems;

CPU: 1000 MHz processor;

Memory: 256Mb of RAM memory.

A.2.2   How to Use

When STROGANOFF is executed, its main screen with three tabs is exhibited (Fig. A.1). From this screen, the basic use of STROGANOFF can be described in three steps:

1. Load the test data file and the training data file;

2. Select the parameters for this experiment;

3. Execute the experiment.

Each of the above steps corresponds to one of the three tabs in the main screen. In the first tab, data files are loaded, and global parameters are set. In the second tab, the specific parameters for each population are defined. In the third tab, the results of the experiments are shown, including the best polynomial tree, the fitness and MDL values, and the predicted values.

In the following subsections we describe how to execute a simple experiment using the STROGANOFF software, according to the above three steps.

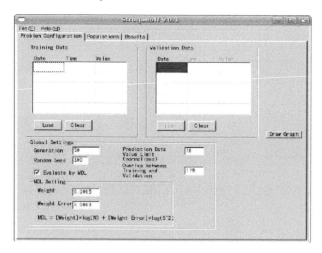

FIGURE A.1: Problem Configuration tab, shown when STROGANOFF starts up.

A.2.3 The General Tab

When you click on the Problem Configuration tab, a screen like the one in Fig. A.1 is displayed. In this tab, you can load the experiment data, and set up global parameters.

A.2.3.1 Loading Data

Two data files are needed to execute an experiment: the Training Data file and the Validation Data file. The Training Data file contains the data that the system will use to build the mathematical model for the Time Series. The Validation Data file contains data that are used to verify that the model built is correct.

In other words, the Training Data file contains the "known" data, and the validation data file contains the "to be predicted" data.

To load the data files, click on the Load button under the Training Data box, and a dialog will prompt you to choose the Training Data file. After that, do the same on the Validation Data box.

You can also edit the data manually by clicking on the numerical value of the data items inside the boxes, and typing the new value using the keyboard.

Finally, after loading both data files, you can click the Draw Graph button, on the right side of the screen. This button shows a graph plotted with the currently loaded data. The Training Data is plotted in red, and the Validation Data is plotted in blue. Use this graph to make sure you loaded the correct data for your experiment.

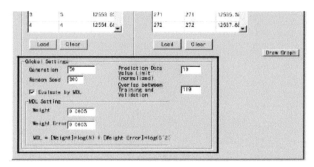

FIGURE A.2: Global Settings area in the Problem Configuration tab.

A.2.3.2 Setting Global Parameters

Global Parameters are those dependent on the experimental data, and cannot be set individually by population. They are located at the bottom box of the screen, which is named global settings (Fig. A.2).

To change the value of a parameter, just click in the value box, and type the new value.

The default values should work for most cases. Pay attention especially to two parameters: the random seed and the Overlap.

The random seed is used to determine all the stochastic events in the experiment. If you want to repeat your experiment sometime later, you will need to remember the value of the random seed. Each value generates a different set of random numbers, which are used to determine crossover, mutation, and other functions based on probability.

The Overlap value determines how many lines in the end of the test data file and in the beginning of the training data file are the same. The files may overlap because the function created by the GP tree has as its terminals past values of the Time Series. If the function is evaluated for the first line of the Training Data, some values from the test data will be required. If the Training Data and the Validation Data are completely different, you may set this value to 0.

For more information about Global parameters, see Section A.2.7.

A.2.4 The Population Tab

The second tab, named *Populations*, is used to set up the GA parameters to one or more populations, which will be used to create a model to the Time Series in the data files. The Population tab is shown on Fig.A.3.

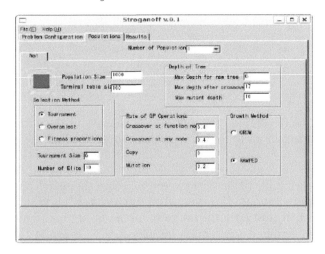

FIGURE A.3: Populations tab, which allows you to set different parameters for each population.

A.2.4.1 Setting up Population Parameters

There are two kinds of parameter fields in this tab: Numerical Fields and Selection Fields.

To change the value of a selection field, click on the radio box next to the desired value. To change the value of a numerical field, simply click on the value box, and type the new value with the keyboard.

There are two special cases regarding the values of numerical fields in this tab: *the tournament size*, and the *Rate of GP Operations*.

- The tournament size is only used with tournament selection. If another selection method is chosen, this box will be grayed out.

- The sum of values of the four parameters inside the Rate of GP Operations should be 1.0 or lower. If the sum is higher, STROGANOFF will show an error screen asking you to fix the parameter's values.

A.2.4.2 Choosing Parameter Values

As with the *Problem Configuration tab*, the default values for the different parameters are good enough to quickly execute an experiment. Keep the following in mind if you wish to change some of these values.

The *population size* parameter determines how many solutions the system will examine at every generation. When this parameter is higher, the search is also broader, but the computational time also rises. Adjust this value based on the complexity of your problem.

The *Terminal table size* indicates how many different variables the system may use to compose the polynomial that models the behavior of the time

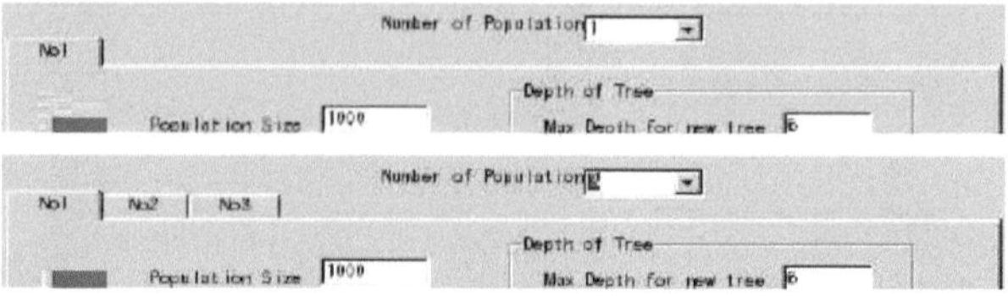

FIGURE A.4: To compare the influence of changes in the parameter values, it is possible to run up to four independent populations at the same time.

series. A larger value of this parameter increases the search space, and may result in better solutions. However, when you increase this value, you decrease the portion of the test data file available for actual testing (a number of lines equal to the Terminal table must be allocated to evaluate the line at the position t_0).

The *Rate of GP Operations* group of values dictate how the search will be performed over the search space. If the mutation ratio is raised, the system will favor exploration of the search space over exploitation. If the values for crossover are higher, it is the other way around. The probability of terminal and non-terminal crossover should be proportional to the size of the Terminal table.

A.2.4.3 Multi Populations

Because the value of the parameters can heavily influence the result of the experiment, STROGANOFF v.0.1 allows up to four independent populations to be run at the same time. In this way, you may test different sets of parameter values, and directly compare how the changes in the parameters influence the final results.

To do this, select the desired value on the Number of Populations dropdown, located at the top of the screen in the Populations tab. When you select a value greater than 1, a number of sub-tabs named "No1", "No2", "No3", and "No4" will appear (see Fig. A.4).

In each of these tabs, you can define the parameter values for each of the populations. The experimental result for each population will be shown separately on the Results tab. On the top-left corner of each tab, there will be a colored rectangle (red, blue, green, or yellow). This color will be used to represent the results of this population in the Results tab.

A.2.5   The Results Tab

The third tab, named Results, is where the experiment is executed, and where the results are displayed. The Results tab is shown in Fig. A.5 in its initial state (before any experiments are executed).

After loading the data, selecting the global parameters and selecting the

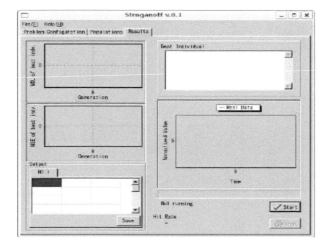

FIGURE A.5: Results tab, which shows information about the experiment.

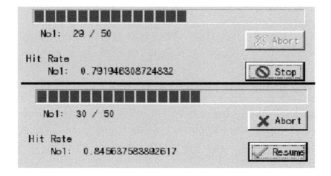

FIGURE A.6: Progress bar and Stop button during execution (above). Abort and Resume buttons with the experiment halted (below).

population parameters, click this tab to begin the experiment.

A.2.5.1 Starting and Stopping an Experiment

To start the experiment, click the button "Start" in the bottom left of the screen. The system will execute the needed calculations, while a progress bar is displayed.

At any time after you start the experiment, you can click on the Stop button to halt it. When halted, press the Resume button to continue the experiment, or the Abort button to cancel it (see Fig. A.6).

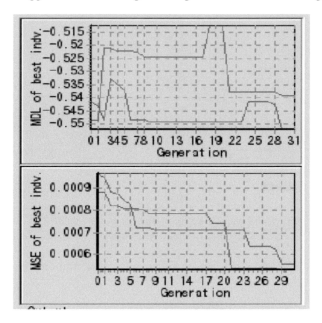

FIGURE A.7: Fitness over time of the best individual in each population.

A.2.5.2   Extracting Evolutionary Results

The two graphs on the top left of the screen show information about the fitness value. During the execution of an experiment, the graphs will be populated with lines that show the fitness value of the best individual of each population over the generations.

Fig. A.7 shows how these two graphs look during an experiment. Each line represents the best individual of one population. The color of the line corresponds to the color of the population, as discussed in Section A.2.4.3.

The upper graph shows the MDL value for the best individual. If the Use MDL fitness option in the Problem Configuration tab is chosen, this value will be used as the fitness.

The lower graph shows the error of the best individual, calculated as the standard deviation of the values predicted by it, and the values in the training data set. If the Use MDL fitness option in the Problem Configuration tab is not chosen, this value will be used as the fitness.

On the top-right side of the tab, a text box will show the tree structure of the current best individual for each population. As the best individual changes over time, the information in this text box is updated. Fig. A.8 shows how the box appears during an experiment.

The best individual for each population is displayed. The structure of the text representation is as follows:

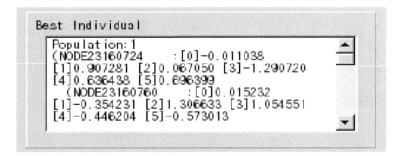

FIGURE A.8: Text representation of the best individuals.

- The representation is Recursive;

- Each terminal is represented as X_n, where n indicates how many time-steps are subtracted to obtain the value of the terminal. For instance, X_5 corresponds to the value $f(t - 5)$ in the time series;

- Each node is represented as:

```
(Node number : [0] a_0 [1] a_1 [2] a_2 [3] a_3 [4] a_4 [5] a_5
      Left sub-node
      Right sub-node)
```

- Node number is the unique identification of that particular node in the program;

- a_0 to a_5 are the constants in the second degree polynomial used in the system:
$$f(x, y) = a_0 + a_1 x + a_2 y + a_3 xy + a_4 x^2 + a_5 y^2 \qquad (A.1)$$

- Left sub-node and right sub-node may be either another node (with the same structure here), or a terminal.

By using the above rules, it is possible to reconstruct the polynomial tree based on the information provided by STROGANOFF.

A.2.5.3 Extracting Prediction Results

On the bottom of the screen, prediction information generated by the best individuals in each population is displayed by the system. This information is displayed by a graph in the lower-right corner, and by a text window on the lower-right (see Fig. A.9).

The text window in the lower-left shows the values predicted by the best individual in the population. The values are normalized. In case more than one

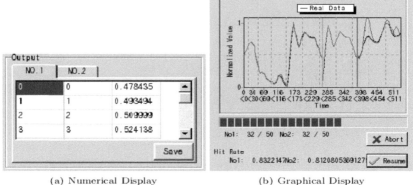

(a) Numerical Display   (b) Graphical Display

FIGURE A.9: Prediction results for an experimental run.

population is being used in the experiment, this window will have one mini-tab for each population to display their results. For example, two populations were used and two tabs can be seen in Fig. A.9.

The predicted values displayed in this text window can be selected with the mouse, and copied for use in other applications.

The graph on the lower right shows a visual representation of the same data displayed in the text window. The black line represents the values from the data files, while the colored lines represent the predicted values from each population. The color of the line corresponds to the color of the rectangles in the Population tabs, as described in Section A.2.4.3.

In the middle of the graph there is a vertical orange line. This line separates the training and validation data sets. The training values are located to the right of the orange line, and the validation values are located to the left.

Also, there may be a vertical purple line in the graph (like in Fig. A.9). This purple line represents validation data which is not used, because it overlaps the training data.

Finally, under the graph and the progress bar, in the extreme lower right there is a numerical value labeled "Hit Rate." The hit rate is defined as the proportion of correct "trend" prediction (correctly predicts that the next value is going to be higher/lower than the current value). The hit-rate is an important accuracy indicator for financial applications.

A.2.6   Data Format

The files used for the Load Training Data and Load Testing Data buttons must be an ASCII file, without special characters. The format of the data inside the file must follow the following pattern.

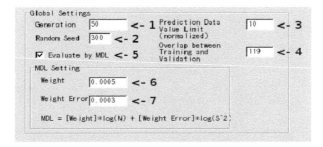

FIGURE A.10: Global parameters.

line number   value date   value
line number   value date   value

...

line number   value date   value

Where line number is the number of the line, value date is the numerical value of the date for that item in the Time Series, and value is the actual value for the Time Series at that point in time. These three values are separated by a tabulation between each of them. Below is an example of the data file bundled together with the program:

145	145	12527.151
146	146	12527.044
147	147	12527.104
148	148	12525.924
149	149	12527.906
150	150	12530.031
151	151	12532.656
152	152	12535.320
153	153	12537.883
154	154	12540.422
155	155	12543.196
156	156	12546.026

A.2.7   List of Parameters

Generation (Parameter 1 in Fig. A.10)

Number of times that the Genetic Algorithm will perform selection and fitness. Must be an integer greater than 0.

Random Seed (Parameter 2)

This number is used to seed the random number generator. Keep the

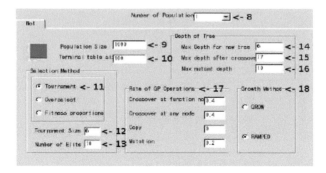

FIGURE A.11: Local parameters.

same random seed to repeat the results of an experiment. Must be an integer.

Prediction Data Value Limit (Parameter 3)

This number limits the maximum predicted value that the program may return. When the Time Series makes a sharp drop or increase, the prediction formula may generate infinite values for those points in time. This parameter bounds those values. Must be a real value. Keep at 10 if you are not sure.

Overlap Between Training and Validation (Parameter 4)

If the Training data set and the Validation data set overlap, you must indicate how many lines overlap in both files here. The program will try to detect this automatically, but may fail if the "date" values in the data set file are not correct. Must be a integer smaller than the total number of lines in either data set.

Evaluate by MDL (Parameter 5)

Check this box if you want to use MDL calculation for selection. Uncheck this box if you only want to use traditional fitness for selection.

Weight (Parameter 6)

This is the weight value for the tree size part of the MDL equation. Must be a real number.

Weight Error (Parameter 7)

This is the weight value for the tree error part of the MDL equation. Must be a real number.

Number of Population (Parameter 8 in Fig. A.11)

This drop-down menu defines how many populations will be run at the same time. The populations are independent, and each may have different values for the following parameters.

Population Size (Parameter 9)

This indicates how many individuals there are in this population. A larger number of individuals may generate a better search, but also takes more time. Must be an integer greater than 0.

Terminal Table Size (Parameter 10)

Indicates how many timesteps in the past the system may use as possible values for terminals in the tree. For instance, if the value of this parameter is 5, the system will use $X(t-1)$, $X(t-2)$, $X(t-3)$, $X(t-4)$, and $X(t-5)$ as possible terminal values. The bigger this parameter is, the larger the search space becomes. Must be an integer larger than 0 and smaller than the number of lines in the training data set and test data set.

Selection Method (Parameter 11)

You may choose one of the three available selection methods. TOURNAMENT uses tournament selection, where some individuals are chosen randomly, and the parent is selected with a chance proportional with the fitness of the individuals in the tournament. OVERSELECT uses a greedy algorithm to generate parent pairs. FITNESS PROPORTIONAL chooses the parent from all individuals at once, with a probability proportional to their fitness.

Tournament Size (Parameter 12)

If the selection method is TOURNAMENT, this parameter defines the size of the tournament. The bigger the tournament, the bigger is the selection pressure. Must be an integer greater than 0 and smaller than the total population.

Number of Elite (Parameter 13)

Every generation, a number of individuals equal to this value is copied directly to the next generation, without suffering crossover or mutation. Must be a positive integer smaller than the total population.

Max Depth for New Tree (Parameter 14) - When creating a new random individual, this value limits the depth of the tree. Must be an integer greater than 0.

Max Depth after Crossover (Parameter 15)

When creating a new individual by crossover, if the resulting tree has depth greater than this value, the crossover is cancelled. Must be an integer greater than 0.

Max Mutant Depth (Parameter 16)

When creating a new individual by mutation, if the resulting tree has depth greater than this value, the mutation is cancelled. Must be an integer greater than 0.

Rate of GP Operations (Parameter 17)

In this box there are four values that determine the relative probability of four different genetic operators. "Crossover at function node" executes the crossover with the crossover point not in the leaf. "Crossover at any node" executes the crossover without restrictions to the crossover point. Copy directly the parents into the new generation. "Mutation" executes the mutation operator (randomly changes one subtree of the individual. The value for each of those fields must be a positive real number, and their sum must be equal to or less than 1.0.

Growth Method (Parameter 18)

Selects a policy when creating a new population. GROW creates all trees randomly, up to the maximum depth. RAMPED creates a variety of large and small trees to avoid the bias towards complexity.

A.3   STROGANOFF for System Identification

Included in this book, together with "STROGANOFF for Time Series Prediction" is also "STROGANOFF for System Identification" (see Section 3.5 for more details). While these two programs are very similar, and use the same engine, there are some key differences between the two.

In this section, we only explain the differences between the sample STROGANOFF program for Time Series prediction and the sample STROGANOFF program for System Identification.

A.3.1   Problem Definition

Both the Time Series Prediction Problem and the System Identification problem consist of developing a model that can output the appropriate response given an input.

However, the model in a Time Series Prediction Problem is based on the past data points. Each new value can be expressed as a function of past values. In the case of System Identification, the inputs and outputs are two different sets of variables. An array of values for the inputs corresponds to one unique result coming from the polynomial tree.

A.3.2   Data Format

The data format for STROGANOFF for System Identification is the first main practical difference between the two programs. Unlike the format described in Section A.2.6, each line contains the variables which refer to one case in the data set.

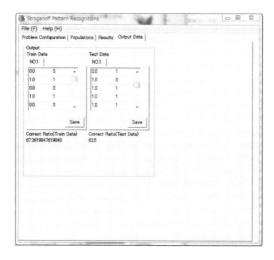

FIGURE A.12: Results tab for the system identification GUI.

The first value of the line is the expected output, followed by one value for each of the input variables in the data. These values are separated by tabulations. As with the Time Series Optimization data format, no special characters are expected in this data file.

A.3.3 Parameters

Two parameters described in Section A.2.7 work differently for System Identification:

- The *Overlap between Training and Validation* parameters has no meaning in System Identification and should not be used.

- The *Terminal Table size* parameter is automatically calculated when the data file is loaded. It shouldn't be manually modified.

A.3.4 Output

STROGANOFF for System Identification adds a fourth tab to the program, which displays the result for each data line in the training and testing data sets (See Fig. A.12).

The tab shows two tables with two columns each. The first table is for the training data, and the second table is for the test data. For each table, the first column shows the output value for each case in the data set, and the second column shows the value calculated by STROGANOFF.

Below each table, a percentage number shows how many correct predictions were made by the algorithm for the training data set and the test data set.

A.4   iGP Library

The source code for the iGP library used in the program (see Section 3.8 for the details of iGP) is included together with the STROGANOFF program. You can use this source code to implement inductive Genetic Programming in your own experiments (for instance, to generate a client to run in a cluster).

In this section we give brief instructions on how to use the library and its main functions in an experiment program. For more detail on the input parameter of the main functions, refer to the comments in the header files.

A.4.1   How to Use

iGP library is written in C++, and can be easily integrated with other programs in this language. To use it, you'll need to compile the library, and call its functions from your program source code.

The source code which accompanies the STROGANOFF package can be compiled under any unix-like system. Just type:

```
% make
```

in the *igplib* sub-directory. To compile it for Macintosh or Windows systems, porting instructions are included in the source code comments.

After compiling, you can use the library by including the object files and header files in your project's source directory, and including the following headers:

- gpc.h

- init.h

- magic.h

- problem.h

- getparams.h

- globals.h

A.4.2   Sample Code

This is an example of a program that uses the basic iGP functions. Use this as a template on how to integrate iGP to your program:

```
//necessary includes:
#include "gpc.h"
#include "init.h"
```

```
#include "magic.h"
#include "problem.h"
#include "getparams.h"
#include "globals.h"

// Loading Datafiles :
   P.inputTrainDataFromFile("train.dat");
   P.inputTestDataFromFile("test.dat");

// Normalizing the Data :
   P.normalize();

// Using the Param object to set
// local and global parameters :
   Param.numgens = 100;

// example of setting a global parameter
   Param.population_size[0] = 100;

// example of setting a local parameter
// - notice the population index ! -
// After setting all the parameters,
// call these functions to initialize them:
   Param.set_global_param();
   Param.set_local_param();

// Initializing global variables :
   init_globals();

// Execute the "run one generation" function
// for as long as needed:
   for (int i=0; i< P.numgens;i++)
      {
        run_one_generation(START_GEN, i, POP, GRID);

        // START_GEN, POP and GRID are global variables,
        // don't need to change these.
        // after each generation you can get information
        // from the training process with these functions.
        Result.update_populations();

        // the parameter is the population index!
        Result.set_best_value(0);
        Result.set_best_pred2(0);
```

```
      // The variable "Result" will have some
      // information about the run.
   }
```

```
// Run the function "free_globals" to free the used memory.
   free_globals();
```

A.4.3   File Contents Summary

- magic.h:

  Defines constants needed by the library

- getparams.c getparams.h:

  Functions for setting and loading global and local parameters.

- globals.c globals.h:

  Global functions, and functions for allocating and freeing them.

- init.c init.h:

  Houses main training loop.

- output.c output.h:

  Functions for outputting the result of the training, and performing predictions.

- problem.c problem.h:

  Functions and definitions for housing the global parameters of the problem.

A.5   MVGPC System Overview

The majority voting genetic programming classifier (MVGPC) has been implemented in Java programming language and named EGPC (ensemble of genetic programming classifiers). The software can be downloaded from the IBA Laboratory homepage (`http://www.iba.t.u-tokyo.ac.jp/english/EGPC.html`). EGPC is a very powerful, easy-to-use Java-based tool for preprocessing and classification of data and for identification of important features from data.

The main features of EGPC are that:

1. It improves the test accuracies of genetic programming rules;

2. It runs in command line interface (CLI) and graphical user interface (GUI) modes;

3. It can be used for binary and multi-class classification;

4. It can handle microarray gene expression data as well as UCI machine learning (ML) databases (with no missing values);

5. It can handle numeric, nominal, and Boolean (converted to numbers) features;

6. It can evolve rules with arithmetic and/or logical functions;

7. It can handle training subsets constructed by fixed or random split of data; and

8. It can handle training and validation data stored in two separate files, provided that they have the same number of features and those features are in the same order in the files.

Moreover, EGPC can be applied repeatedly to the data for a reduced set of features and possibly for better test accuracies; in this case, EGPC acts as a preprocessor of features.

A.5.1 Data Format

EGPC can handle tab, colon, semicolon, or space delimited text files. The files can be in either microarray or UCI ML datafile format. In microarray format, the first row contains the class labels of the instances, and the other rows contain the gene expression values of different genes in different instances. One file: *BreastCancer.txt* in the examples is in this format.

In UCI ML format, the first column contains the class labels of the instances, and the other columns contain the values of attributes in different instances. Two files: *Monk.txt* and *WCBreast.txt* in the examples are in this format.

In either format, the class labels of instances must be numeric values starting from 0, i.e., for binary classification, the class labels of the instances will be either 0 or 1 while for multi-class classification, the class labels will be from 0 to $(c - 1)$, where c is the number of classes in the data file.

A.5.2 Training and Test Data Constructions

The training and the test data can be in one file or in two separate files. If the data are in two separate files, the split of the whole data is fixed. If the data are in one file, the training and test subsets can be constructed randomly or by fixed split of the data. In the case of random split, the information about training subset is entered by combining the training sizes of different classes, delimited by colon (:), into a single string. Let us give an example of this. Suppose that a data set consists of three classes of instances, and the number of instances of the classes is 8, 8, and 6, respectively. If the training subset consists of 50% of the instances, the string containing information about the training data would be 4:4:3. However, in the case of the fixed split of the instances stored in one file, a file containing the indexes of the training instances should be passed to the software. See *BreastTrainIndex.txt* for an example of the fixed split of *BreastCancer.txt* data. Note that the index of the first instance is 1.

A.5.3 Features (Attributes/Genes)

EGPC can handle numeric, nominal and Boolean attributes. If any feature is in nominal format, it must be converted into numeric values. For example, if the possible values of a nominal feature is Sunny, Cloudy, Rainy, and Snowing, these values should be converted to 0, 1, 2, and 3, respectively. The identifiers for the three types of features are as follows: N: numeric; L: nominal; and B: Boolean. If a range of the features are of the same type, it can be indicated by *<start index>:<end index>*. For example, if the first 30 features of a data set are numeric, it can be indicated by *N1:30*. For nominal features, there is one extra parameter, the maximum value of the nominal features, which follows the *<end index>* separated by colon. An example is *L1:30:4* where the first 30 features are nominal and have the maximum value of 4. For multi-type features, each type and range should be indicated by the above notations and separated by colon. An example of this notation is *N1:30:L31:50:4:B51:60*, in which the first 30 features are numeric, the next 20 features are nominal and have the maximum value of 4, and the last 10 features are binary.

Each feature in the data file is represented by an '*X*' followed by the feature number. For example, *X3* represents the third feature of a data set.

A.5.4 Functions

EGPC can handle the following functions :{ $+$, $-$, $*$, $/$, SQR, SQRT, LN, EXP, SIN, COS, AND, OR, NOT, $=$, $<>$, $>$, $<$, $>=$, $<=$}. If all the features are numeric, one can use arithmetic functions: {$+$, $-$, $*$, $/$, SQR, SQRT, LN, EXP, SIN, COS}, or logical and Boolean functions: {AND, OR, NOT, $=$, $<>$, $>$, $<$, $>=$, $<=$}. Note that since all the features are numeric, only AND, OR, NOT cannot be used. If all the features are either nominal or Boolean, we recommend using logical and Boolean functions only. However, EGPC would be able to handle the combinations as nominal features are converted to numeric values (by the user) before running the software.

Some functions are executed in protected mode to avoid undefined results—underflow or overflow. These functions are treated as follows:

- $SQRT(Y) = 0$ if Y is negative;
- $Y/Z = 1$ if Z is 0;

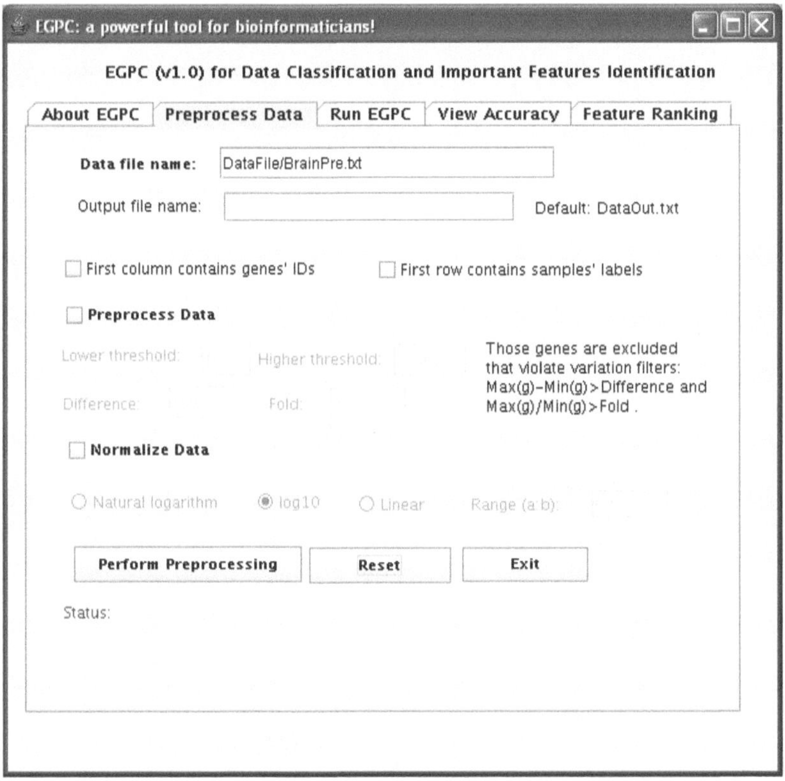

FIGURE A.13: A screen shot of GUI of EGPC (*Preprocess Data* tab).

- $EXP(Y) = EXP(Max(-10000, Min(Y, 20)))$; Y is bounded in $[-10000, 20]$;

- $LN(Y) = 0$ if $Y = 0$; otherwise $LN(Y) = LN(|Y|)$ and Y is bounded in $[-1.0e+100, 1.0e+100]$; and

- $SIN(Y)$ and $COS(Y)$: Y is bounded in $[-10000, 10000]$.

A.5.5 Fitness Evaluation

Assuming that the distribution of the classes in a data set is balanced, a GP rule is evaluated using correlation based fitness function defined by equation (4.26).

FIGURE A.14: A screen shot of GUI of EGPC (*Run EGPC* tab).

A.5.6   Ensemble Size

The number of single rules or sets of rules that participate in majority voting is indicated here as *ensemble size*. The minimum ensemble size is three. For binary classification, we recommend an odd value for the ensemble size.

A.5.7   Evolved Rules (S-expressions)

The expression of a rule is called S-expression. The predicted class of a rule is determined depending on the output of the rule:

- Boolean output (an S-expression consists of logical only or logical and arithmetic functions):

  - Binary classification: IF (S-expression is true) THEN 0 ELSE 1.

  - Multi-class classification: IF (S-expression is true) THEN Target-Class ELSE Other.

Real-valued output (an S-expression consists of arithmetic functions only):

- – Binary classification: IF (S-expression $\geq$ 0) THEN 0 ELSE 1.
- – Multi-class classification: IF (S-expression $\geq$ 0) THEN TargetClass ELSE Other.

Therefore, the slice point for a real-valued output is 0.

A.5.8 Default Settings of GP Parameters

The default settings of some GP parameters in EGPC are as follows:

- Maximum number of nodes in a GP tree $= 100$;

- Maximum initial depth $= 6$;

- Initial population generation method: ramped half-and-half;

- Maximum crossover depth $= 7$;

- Selection method for crossover: greedy-over;

- Reproduction probability $= 0.1$;

- Crossover probability $= 0.9$;

- Mutation probability $= 0.1$;

- Termination criteria: fitness $= 1.0$ or maximum number of generations has passed; and

- Regeneration type: elitism (elite size $= 1$).

A.5.9 Example Files Included with the Software Bundle

A.5.9.1 Monk's Problem (*Monk.txt*):

Monk's Problem has been downloaded from `http://www.ics.uci.edu/~mlearn/MLRepository.html`. It is a binary classification problem, and consists of 6 nominal features (values: 1–4), and a total of 556 instances divided into 124 training and 432 test instances. We have put the 432 independent test instances into the file *MonkValid.txt* that are used as a validation file for the problem. The data are in UCI ML format. The goal task is simple; i.e., the target concept is: IF ((X1 = X2) OR (X5 = 1)) THEN 1 ELSE 0.

A.5.9.2 Wisconsin Breast Cancer Data (*WCBreast.txt*):

The file has been downloaded from `http://www.ics.uci.edu/~mlearn/MLRepository.html`. It consists of 30 numeric features, and 569 instances. It is a binary classification problem. To use this data set as an example, we randomly split it into training and test subsets with 1:1 ratio.

A.5.9.3 Breast Cancer Data (*BreastCancer.txt*):

The file is a microarray data consisting of the gene expression values of 3226 preprocessed genes across 22 instances. It is a three-class classification problem. The names of the classes are BRAC1, BRAC2, or Sporadic. All the features are numeric. The training subset consists of the fixed split of this data set, and the indexes of the training instances are stored in the file *BreastTrainIndex.txt*. The number of training and test instances is 17 and 5, respectively. The number of BRAC1, BRAC2, and Sporadic instances in the training and test subsets are 6, 6, 5, and 2, 2, 1, respectively. The original data set is available at `http://research.nhgri.nih.gov/microarray/NEJM_Supplement/`.

A.5.9.4 Brain Cancer Data (*BrainPre.txt*):

The file is a microarray data file consisting of expression values of 12625 genes across 50 instances. It is a binary classification problem. This data file is provided as an example file for data preprocessing. The original data file is available at `http://www-genome.wi.mit.edu/cancer/pub/glioma`.

A.5.10 Execution of the Software

The EGPC software is implemented in Java programming language and available as a jar (Java Archive) executable file. Three jar files: *EGPCpre.jar*, *EGPCcom.jar*, and *EGPCgui.jar* are included with this software bundle. *EGPCpre.jar* is for preprocessing of microarray data in CLI mode; *EGPCcom.jar* and *EGPCgui.jar* are for data classification and important feature identification in CLI and GUI modes, respectively. *EGPCgui.jar* also has interfaces for data preprocessing in GUI mode.

Given an input file for preprocessing, *EGPCpre.jar* will create two output files containing the preprocessed data and the cross-reference indexes of the features. If the name of the output file is not provided, the preprocessed data will be in the file *DataOut.txt*; the file containing cross-reference indexes is *CrossRefIdx.txt*.

EGPCcom.jar and *EGPCgui.jar* will create three files containing rules, accuracy, and feature frequency information in the working directory (from where the software is run). If the name of a data file is *Example.txt*, the software will create three files named *RulesExample.txt*, *AccuracyExample.txt*, and *GeneFreqExample.txt*.

A.5.10.1   Execution of *EGPCpre.jar* in CLI Mode

To run the program, type:

```
java [-Xmx<heap size>] -jar EGPCpre.jar [arguments...]
```

at the command prompt. Command-line arguments and formats are as follows:

- -Xmx<heap size>: maximum heap size; some data sets may require higher heap size. Example: -Xmx512m (m or M for megabyte).

- -f <input file>: input data file name (with path if not on the current working directory); it must be provided.

- -o <output file>: output file name (with path if not on the current working directory); default: DataOut.txt.

- -p <l:h:d:f>: preprocessing parameters; l = lower threshold, h = higher threshold, d = difference, f = fold change.

- -n <normalization info>: normalization info; for log normalization type G with the base like G10 or Ge while for linear normalization type La:b where a:b is the range.

- -h <header info>: header info; G: first column contains headers; S: first row contains headers; GS or SG for both.

One example of operation is given below.

```
java -jar EGPCpre.jar -f "DataFile/BrainPre.txt" -o BrainPro.
txt -p 20:16000:100:3 -n Ge -h GS
```

A.5.10.2   Execution of *EGPCcom.jar* in CLI Mode:

To run the program, type:

```
java [-Xmx<heapsize>] -jar EGPCcom.jar [arguments...]
```

at the command prompt. Command line arguments and formats are as follows:

- -Xmx<heap size>: maximum heap size; some data sets may require higher heap size. Example: -Xmx512m (m or M for megabyte).

- -u: UCIML format; default (if it is omitted) is microarray format.

- -d <data file>: data file name (with path if not on the current working directory); it must be provided.

- -v <validation file>: validation file name (with path if not on the current working directory); if it is not provided, the training information must be provided under the "-t" command line argument below.

- -s <instance size>: number of instances; it must be provided.

- -a <feature size>: number of features; it must be provided.

- -A <feature info>: feature information; default is that all features are numeric. See above "Features" for details about feature information.

- -t <training info>: training subset information; the training information can be either the file name (with path if not on the current working directory) containing the indexes of the training instances or the training size of each type of instance delimited by colons like 179:106.

- -c <classes>: number of classes; default is 2.

- -m <ensemble size>: ensemble size; default is 3.

- -F <functions>: functions to be used; functions are delimited by colon (:) and the default functions are "+:−:/:*:sqr:sqrt". Note here that the functions' string must be within double quotation (" ").

- -p <population size>: population size; default is 1000.

- -g <max gen>: maximum number of generations; default is 50.

- -r <max run>: number of trials or repetitions; default is 20.

A.5.10.3 Execution of *EGPCcom.jar* on the First Three Example Files from Command Prompt:

- Monk problem:

```
java -jar EGPCcom.jar -u -d "DataFile/Monk.txt" -v
"DataFile/MonkValid.txt" -s 556 -a 6 -A L1:6:4 -F
">:<:=:<>:AND:OR:NOT:>=:<=" -m 3 -c 2 -r 3
```

- Wisconsin breast cancer data:

```
java -jar EGPCcom.jar -u -d "DataFile/WCBreast.txt" -s 569
-t 179:106 -a 30 -A N1:30 -F "+:-:*:/:SQRT:SQR:>:<:=:<>:
AND:OR:NOT:>=:<=" -m 3 -c 2 -r 3
```

- Breast cancer data:

```
java -jar EGPCcom.jar -d "DataFile/BreastCancer.txt" -s 22
-a 4434 -t "DataFile/BreastTrainIndex.txt" -A N1:4434 -F
"+:-:/:*:SQRT:SQR" -m 3 -c 3 -r 3
```

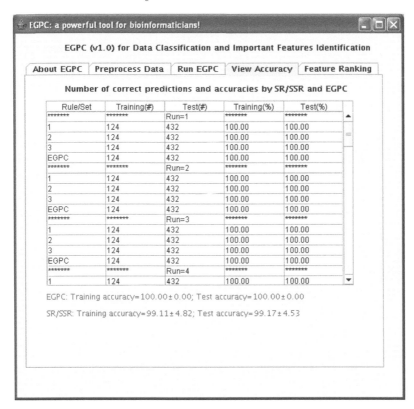

FIGURE A.15: A screen shot of GUI of EGPC (*View Accuracy* tab).

A.5.10.4 Execution of *EGPCgui.jar* in GUI mode:

To run the program, type the following command

```
java [-Xmx<heap size>] -jar EGPCgui.jar
```

at the command prompt.

Snapshots of the software in GUI mode are shown in Figs. A.13–A.16. In the GUI view, the first tab is about the software. *Preprocess Data* is for preprocessing of microarray gene expression data. The data file name and values of different parameters are entered from *Run EGPC* tab. While EGPC is being executed on a data file, the number of correct predictions and the accuracies by single rules or sets of rules, as well as those by the EGPC, can be viewed on *View Accuracy* tab. The more frequently selected features are displayed in the tab *Feature Ranking*. Here also three output files for rules, accuracy, and feature frequency are created.

The previously mentioned three examples can also be run in graphical-user interface mode.

FIGURE A.16: A screen shot of GUI of EGPC (*Feature Ranking* tab).

A.6   EDA Simulator

In this simulator, the maximum X value for the given function within a certain range of values for X is found, using various versions of EDAs.

The detailed instructions on how to use the program are as follows:

1. Initial Settings
   Before searching for the maximum value, you need to enter the following configuration.

   - Set function
     Enter here the function to be used. The function must be written in Excel format.

   - Set x range
     Enter here the range for the x variable to be maximized.   The

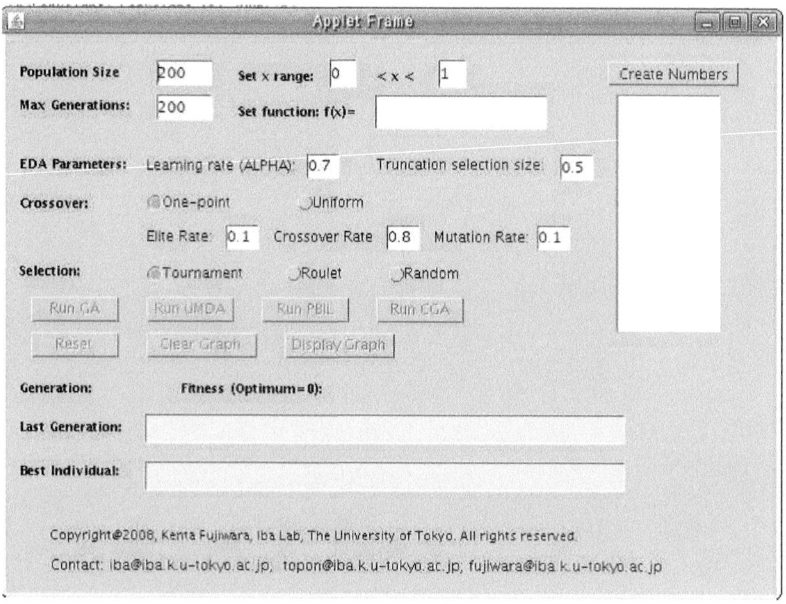

FIGURE A.17: A screen shot of EDA simulator.

program will find the x value which maximizes the function in this range.

- Population size
  Enter here the number of individuals for the Genetic Algorithm.

- Max generations
  Enter here the maximum number of iterations for the program. The genetic algorithm will run this number of generations.

After finishing the above configuration, click the button "Create Numbers." It will randomly choose numbers from the defined interval as the initial population. At this time, a graph of the defined function will be shown.

2. Selecting the search heuristic
   Select the Evolutionary Algorithm you wish to use to search for the maximum value of the function. You can try and compare the results of several different heuristics.

   - Run GA
     Search the optimal value using Genetic Algorithms.

   - Run UMDA
     Search the optimal value using UMDA.

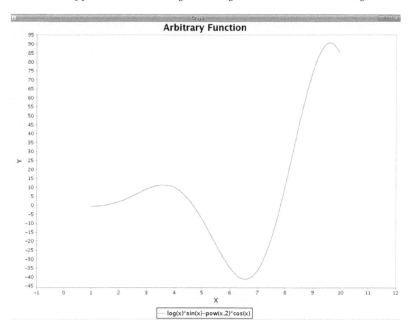

FIGURE A.18: An example of objective function defined by "Set function" command.

- Run PBIL
  Search the optimal value using PBIL.

- Run CGA
  Search the optimal value using CGA.

3. Results
   After the program finishes calculating, a graph is displayed. The graph displays the average value of the function for every generation.

   - Reset
     Returns to the initial settings.

   - Clear Graph
     Removes the display of the Graph.

   - Display Graph
     Returns the display of the Graph.

   - Last Generation
     Displays the individuals in the last generation.

   - Best Individual
     Displays the best individual of each generation.

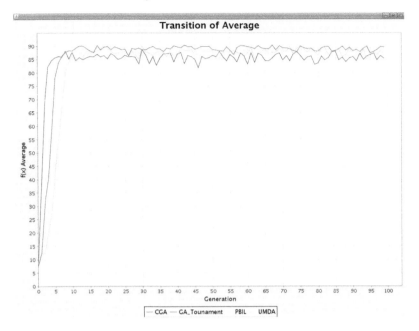

FIGURE A.19: Comparison of EDA and GA performance results.

4. Parameters
   Sets the parameters for each search heuristic.

   - Learning rate
     Learning parameter for the PBIL heuristic.

   - Truncation selection size
     Selects individuals with fitness value above this.

   - Crossover
     Selects the crossover method. One-point Crossover and Uniform crossover are available.

   - Elite Rate
     Proportion of individuals to copy to the next generation, according to the Elite strategy.

   - Crossover Rate
     Probability of using the Crossover operator.

   - Mutation Rate
     Probability of using the Mutation operator.

   - Selection
     Selects the selection method. Tournament, Roulette, and Random are available.

References

[Abraham and Grosan06] Abraham, A. and Grosan, C., Decision support systems using ensemble genetic programming, *Journal of Information & Knowledge Management*, vol. 5, no. 4, pp. 303–313, 2006.

[Aha and Bankert95] Aha, D. W. and Bankert, R. L., A comparative evaluation of sequential feature selection algorithms, in *Proc. of the Fifth International Workshop on Artificial Intelligence and Statistics*, pp. 1–7, Springer-Verlag, 1995.

[Aha *et al.*91] Aha, D. W., Kibler, D., and Albert, M., Instance-based learning algorithms, *Machine Learning*, vol. 6, pp. 37–66, 1991.

[Alizadeh *et al.*00] Alizadeh, A., Eisen, M., Davis, R., Ma, C., Lossos, I., Rosenwald, A., Boldrick, J., Sabet, H., Tran, T., Yu, X., Powell, J., Yang, L., Marti, G., Moore, T., Hudson, J. J., Lu, L., Lewis, D., Tibshirani, R., Sherlock, G., Chan, W., Greiner, T., Weisenburger, D., Armitage, J., Warnke, R., Levy, R., Wilson, W., Grever, M., Byrd, J., Botstein, D., Brown, P., and Staudt, L., Distinct types of diffuse large B-cell lymphoma identified by gene expression profiling, *Nature*, vol. 403, no. 6781, pp. 503–511, 2000.

[Alon *et al.*99] Alon, U., Barkai, N., Notterman, D. A., Gish, K., Ybarra, S., Mack, D., and Levine, A. J., Broad patterns of gene expression revealed by clustering of tumor and normal colon tissues probed by oligonucleotide arrays, in *Proc. of National Academy of Science, Cell Biology*, vol. 96, pp. 6745–6750, 1999.

[Ando and Iba04] Ando, S. and Iba, H., Classification of gene expression profile using combinatory method of evolutionary computation and machine learning, *Genetic Programming and Evolvable Machines*, vol. 5, pp. 145–156, 2004.

[Ando and Iba05] Ando, D. and Iba, H., Real-time Musical interaction between musician and multi-agent system, in *Proc. of the 8th Generative Art Conference (GA 2005)*, pp. 93–100, 2005.

[Angeline96] Angeline, P., Two self-adaptive crossover operators for genetic programming, *Advances in Genetic Programming 2*, Angeline, P. and Kinnear, K. (eds.), MIT Press, 1996.

[Angeline *et al.*94] Angeline, P., Saunders, G. M., and Pollack, J. B., An evolutionary algorithm that constructs recurrent neural networks, *IEEE Transactions Neural Networks*, vol. 5, no. 1, pp. 54–65, 1994.

[Aranha and Iba08] Aranha, C. and Iba, H., A tree-based GA representation for the portfolio optimization problem, in *Proc. of the 10th Genetic and Evolutionary Computation Conference (GECCO2008)*, pp. 873–880, ACM Press, 2008.

[Aranha *et al.*07] Aranha, C., Kasai, O., Uchide, U., and Iba, H., Day-trading rules development by genetic programming, in *Proc. of the 6th International Conference on Computational Intelligence in Economics & Finance (CIEF)*, pp. 515–521, 2007.

[Armstrong91] Armstrong, W. W., Learning and generalization in adaptive logic networks, *Artificial Neural Networks*, Kohonen, T. (ed.), pp. 1173–1176, Elsevier Science, 1991.

[Astrom and Eykhoff71] Astrom, K. J. and Eykhoff, P., System identification, a survey, *Automatica*, vol. 7, pp. 123–162, 1971.

[Atkeson *et al.*97] Atkeson, C. G., Moore, A. W., and Schaal, S., Locally weighted learning, *Artificial Intelligence Review*, vol. 11, pp. 11–73, 1997.

[Attias99] Attias.,H., Inferring parameters and structure of latent variable models by variational Bayes, in *Proc. of the 15th Conference of Uncertainty in Artificial Intelligence*, pp. 21–30, Stockholm, Sweden, Morgan Kaufmann, 1999.

[Au *et al.*04] Au, N., Gown, A., Cheang, M., Huntsman, D., Yorida, E., Elliott, W. M., Flint, J., English, J., Gilks, C., and Grimes, H., p63 expression in lung carcinoma: A tissue microarray study of 408 cases, *Applied Immunohistochemistry & Molecular Morphology*, vol. 12, no. 3, pp. 240–247, 2004.

[Bäck96] Bäck,T., *Evolutionary Algorithms in Theory and Practice*, Oxford University Press, 1996.

[Bäck *et al*.00] Back, T., Fogel, D. B., and Michalewicz, Z., *Evolutionary Computation 1: Basic Algorithms and Operators*, Taylor & Francis, 2000.

[Baluja94] Baluja, S., Population-based incremental learning: A method for integrating genetic search based function optimization and competitive learning, Technical Report CMU-CS-94-163, Pittsburgh, PA, 1994.

[Banzhaf *et al*.98] Banzhaf, W., Nordin, P., Keller, R. E., and Francone, F. D., *Genetic Programming: An Introduction. On the Automatic Evolution of Computer Programs and Its Applications*, Morgan Kaufmann, 1998.

[Barandela *et al*.04] Barandela, R., Valdovinos, R. M., Sánchez, J. S., and Ferri, F. J., The imbalanced training sample problem: Under or over sampling?, in *Proc. of Joint IAPR international workshops SSPR 2004 and SPR 2004*, vol. 3138 of *LNCS*, pp. 806, Lisbon, Portugal, Springer-Verlag, 2004.

[Barzdins and Barzdins91] Barzdins, J. M. and Barzdins, G. J., Rapid construction of algebraic axioms from samples, *Theoretical Computer Science*, vol. 90, pp. 179–208, 1991.

[Belew *et al*.91] Belew,R.K., McInerney, J., and Schraudolph, N. N., Evolving networks: Using genetic algorithm with connectionist learning, *Artificial Life II*, Langton, C. G. et al. (eds.), Addison-Wesley, 1991.

[Ben-Dor *et al*.00] Ben-Dor, A., Bruhn, L., Friedman, N., Nachman, I., Schummer, M., and Yakhini, Z., Tissue classification with gene expression profiles, *Journal of Computational Biology*, vol. 7, pp. 559–584, 2000.

[Ben-Dor *et al*.99] Ben-Dor, A., Shamir, R., and Yakhini, Z., Clustering gene expression patterns, *Journal of Computational Biology*, vol. 6, pp. 281–297, 1999.

[Bhattacharjee *et al*.01] Bhattacharjee, A., Richards, W., Stauton, J., Li, C., Monti, S., Vasa, P., Ladd, C., Behesti, J.,

Buneo, R., Gillete, M., Loda, M., Weber, G., Mark, E., Lander, E., Wong, W., Johnson, B., Golub, T., Sugarbaker, D., and Meyerson, M., Classification of human lung carcinomas by mRNA expression profiling reveals distinct adenocarcinoma subclasses, *Proceedings of National Academy of Science*, vol. 98, pp. 13790–13795, 2001.

[Bishop95] Bishop,C., *Neural Networks for Pattern Recognition*, Oxford University Press, Oxford, UK, 1995.

[Box and Jenkins70] Box, G. E. P. and Jenkins, G. M., *Time Series Analysis Forecasting and Control*, Holden-Day, San Francisco, CA, 1970.

[Breiman96] Breiman, L., Bagging predictors, *Machine Learning*, vol. 24, no. 2, pp. 123–140, 1996.

[Breiman01] Breiman, L., Random forests, *Machine Learning*, vol. 45, no. 1, pp. 5–32, 2001.

[Bruzzone and Serpico97] Bruzzone, L. and Serpico, S. B., Classification of imbalanced remote-sensing data by neural networks, *Pattern Recognition Letters*, vol. 18, no. 11–13, pp. 1323–1328, 1997.

[Chang and Lin01] Chang, C. C. and Lin, C. J., LIBSVM: A library for support vector machines, Software available at http://www.csie.ntu.edu.tw/~cjlin/libsvm, 2001.

[Chawla *et al.*02] Chawla, N. V., Bowyer, K. W., and Kegelmeyer, W. P., Smote: Synthetic minority over-sampling technique, *Journal of Artificial Intelligence Research*, vol. 16, pp. 321–357, 2002.

[Chen *et al.*98] Chen, S. H., Yeh, C. H., and Lee, W. C., Option pricing with genetic programming, in *Genetic Programming 1998: Proc. of the 3rd Annual Conference*, University of Wisconsin, Madison, Wisconsin, 22-25 July, pp. 22–27, Morgan Kaufmann, 1998.

[Chidambaran *et al.*98] Chidambaran, N. K., Lee, C. H. J., and Trigueros, J. R., An adaptive evolutionary approach to option pricing via genetic programming, in *Genetic Programming 1998: Proc. of the 3rd Annual Conference*, University of Wisconsin, Madison, Wisconsin, 22-25 July, pp. 38–41, Morgan Kaufmann, 1998.

[Chien *et al.*02] Chien, B. C., Lin, J., and Hong, T. P., Learning discriminant functions with fuzzy attributes for classification using genetic programming, *Expert Systems with Applications*, vol. 23, pp. 31–37, 2002.

[Cooper and Herskovits92] Cooper, G. and Herskovits, E., A Bayesian method for the induction of probabilistic networks from data, *Machine Learning*, vol. 9, pp. 309–347, 1992.

[Dasarathy91] Dasarathy, B., *Nearest neighbor (NN) norms: NN pattern classification Techniques*, IEEE Computer Society Press, 1991.

[Deb and Goldberg92] Deb, K. and Goldberg, D.E., Analyzing deception in trap functions, *Foundations of Genetic Algorithms 2*, Whitley, L. D. (ed.), pp. 93–108, Morgan Kaufmann, 1992.

[Deb and Reddy03] Deb, K. and Reddy, A. R., Reliable classification of two-class cancer data using evolutionary algorithms, *BioSystems*, vol. 72, pp. 111–129, 2003.

[de Jong and Bosman04] de Jong, E. D. and Bosman, P. A. N., Grammar transformations in an EDA for genetic programming, Technical Report UU-CS-2004-047, Institute of Information and Computing Sciences, Utrecht University, 2004.

[de Menezes and Nikolaev06] de Menezes, L. and Nikolaev, N., Forecasting with genetically programmed polynomial neural networks, *International Journal of Forecasting*, vol. 22, pp. 249-265, 2006.

[Deutsch03] Deutsch, J. M., Evolutionary algorithms for finding optimal gene sets in microarray prediction, *Bioinformatics*, vol. 19, no. 1, pp. 45–52, 2003.

[Ding00] Ding, C., Tumor tissue classification using support vector machines and k-nearest neighbor method, in *Proc. of the 1st Conference on Critical Assessment of Microarray Data Analysis (CAMDA2000)*, Duke University, Durham, N.C., 18-19 December, 2000

[Ding03] Ding, C. H. Q., Unsupervised feature selection via two-way ordering in gene expression analysis, *Bioinformatics*, vol. 19, no. 10, pp. 1259–1266, 2003.

[Domingos99] Domingos, P., Metacost: A general method for making classifiers cost-sensitive, in *Proc. the*

5th *ACM SIGKDD International Conference on Knowledge Discovery and Data Mining*, pp. 155–164, San Diego, CA, 1999.

[Driscoll *et al.*03] Driscoll, J. A., Worzel, B., and MacLean, D., Classification of gene expression data with genetic programming, *Genetic Programming Theory and Practice*, Riolo, R. L. and Worzel, B. (eds.), pp. 25–42, Kluwer Academic Publishers, 2003.

[Dudoit *et al.*02] Dudoit, S., Fridlyand, J., and Speed, T. P., Comparison of discrimination methods for the classification of tumors using gene expression data, *Journal of the American Statistical Association*, vol. 97, no. 457, pp. 77–87, 2002.

[Eckman *et al.*03] Eckman, E., Watson, M., Marlow, L., Sambamurti, K., and Eckman, C. B., Alzheimer's disease beta-amyloid peptide is increased in mice deficient in endothelin-converting enzyme, *Journal of Biological Chemistry*, vol. 278, no. 4, pp. 2081–2084, 2003.

[Eiben and Smith03] Eiben, A. E. and Smith, J. E., *Introduction to Evolutionary Computing*, Springer-Verlag, 2003.

[Eshelman91] Eshelman, L., The CHC adaptive search algorithm: How to have safe search when engaging in nontraditional genetic recombination, *Foundations of Genetic Algorithms I*, Rawlins, G. J. E. (ed.), pp. 265–283, Morgan Kauffmann, 1991.

[Falco *et al.*02] Falco, I. D., Cioppa, A. D., and Tarantino, E., Discovering interesting classification rules with genetic programming, *Applied Soft Computing*, vol. 1, pp. 257–269, 2002.

[Faraway and Chatfield98] Faraway, J. and Chatfield, C., Time series forecasting with neural networks: A comparative study using the airline data, *Applied Statistics*, vol. 47, no. 2, pp. 231–250, 1998.

[Farlow84] Farlow, S.J. (ed.), *Self-Organizing Methods in Modeling, GMDH Type Algorithms*, Marcel Dekker, Inc., 1984.

[Fisher36] Fisher, R., The use of multiple measurements in taxonomic problems, *Ann. Eugenics*, vol. 7, 1936.

[Fogel93]   Fogel, D. B.,   Evolving behaviors in the iterated prisoner's dilemma,   *Evolutionary Computation*, vol. 1, no. 1, 1993.

[Fogel99]   Fogel, D. B., *Evolutionary Computation: Toward a New Philosophy of Machine Intelligence*, IEEE Press, 1999.

[Fogel *et al.*66]   Fogel, L. J, Owens, A. J., and Walsh, M. J., *Artificial Intelligence through Simulated Evolution*, Wiley, 1966.

[Folino *et al.*06a]   Folino, G., Pizzuti, C., and Spezzano, G., GP ensembles for large-scale data classification, *IEEE Transactions on Evolutionary Computation*, vol. 10, no. 5, pp. 604–616, 2006.

[Folino *et al.*06b]   Folino, G., Pizzuti, C., and Spezzano, G., Improving cooperative GP ensemble with clustering and pruning for pattern classification, in *Proc. of the 8th Genetic and Evolutionary Computation Conference (GECCO2006)*, pp. 791–798, ACM Press, 2006.

[Franke82]   Franke, R., Scattered data interpolation: Tests of some methods, *Mathematics of Computation*, vol. 38, pp. 181–200, 1982.

[Freund and Schapire97]   Freund, Y. and Schapire, R. E.,   A decision-theoretic generalization of on-line learning and an application to boosting, *Journal of Computer and System Sciences*, vol. 55, no. 1, pp. 119–139, 1997.

[Furey *et al.*00]   Furey, T., Cristianini, N., Duffy, N., Bednarski, D. W., Schummer, M., and Haussler, D., Support vector machine classification and validation of cancer tissue samples using microarray expression data, *Bioinformatics*, vol. 16, no. 10, pp. 906–914, 2000.

[Gathercole and Ross96]   Gathercole, C. and Ross, P., An adverse interaction between crossover and restricted tree depth in genetic programming, in *Genetic Programming 1996: Proceedings of the First Annual Conference*, Koza, J., Goldberg, D. E., Fogel, D. B., and Riolo, R. L. (eds.), Stanford University, CA, 28-31 July, pp. 291–296, MIT Press, 1996.

[Giles *et al.*92]   Giles, C. L., Miller, C. B., Chen, D., Chen, H. H., Sun, G. Z., and Lee, Y. C., Learning and extracting finite state automata with second-order recurrent neural networks, *Neural Computation*, vol.4, no. 3, pp. 393–405, 1992.

[Glover89]   Glover, F., Tabu search, *ORSA Journal on Computing*, vol. 1, no. 3, pp. 190–206, 1989.

[Goldberg89]   Goldberg, D. E., *Genetic Algorithms in Search, Optimization and Machine Learning*, Addison-Wesley Pub., 1989.

[Golub *et al.*99]   Golub, T., Slonim, D., Tamayo, P., Huard, C., Gaasenbeek, M., Mesirov, J., Coller, H., Loh, M., Downing, J., Caligiuri, M., Bloomfield, C., and Lander, E., Molecular classification of cancer: Class discovery and class prediction by gene expression monitoring, *Science*, vol. 286, no. 15, pp. 531–537, 1999.

[Gorman and Sejnowski88]   Gorman, R. P. and Sejnowski, T. J., Analysis of hidden units in a layered network trained to classify sonar targets, *Neural Networks*, vol. 1, 1988.

[Grosan and Abraham06]   Grosan, C. and Abraham, A., Stock market modeling using genetic programming ensembles, *Genetic Systems Programming: Theory and Experiences*, Nedjah, N., Abraham, A., and de Macedo Mourelle, L. (eds.), pp. 133–148, Springer, 2006.

[Guyon *et al.*02]   Guyon, I., Weston, J., Barnhill, S., and Vapnik, V., Gene selection for cancer classification using support vector machine, *Machine Learning*, vol. 46, no. 1–3, pp. 389–422, 2002.

[Han *et al.*05]   Han, H., Wang, W., and Mao, B. H., Borderline-SMOTE: A new over-sampling method in imbalanced data sets learning, in *Proc. of the International Conference on Intelligent Computing*, vol. 3644 of *LNCS*, pp. 878–887, Springer-Verlag, 2005.

[Hanley and Mcneil82]   Hanley, J. A. and Mcneil, B. J., The meaning and use of the area under a receiver operating characteristic curve, *Radiology*, vol. 143, no. 1, pp. 29–36, 1982.

[Harik99]   Harik, G., Linkage learning via probabilistic modeling in the ECGA, Technical Report, Illigal Report 99010, Illinois Genetic Algorithms Laboratory, University of Illinois at Urbana-Champaign, 1999.

[Hasegawa and Iba06] Hasegawa, Y. and Iba, H., Estimation of Bayesian network for program generation. in *Proc. of the 3rd Asian-Pacific Workshop on Genetic Programming*, pp. 35–46, Hanoi, Vietnam, 2006.

[Hasegawa and Iba08] Hasegawa, Y. and Iba, H., A Bayesian network approach for program generation, *IEEE Transactions on Evolutionary Computation*, vol. 12, no. 6, pp. 750–764, 2008.

[Hasegawa and Iba09] Hasegawa, Y. and Iba, H., Latent variable model for estimation of distribution algorithm based on probabilistic context-free grammar (accepted for publication), *IEEE Transactions on Evolutionary Computation*, 2009.

[Heckerman95] Heckerman, D., A tutorial on learning with Bayesian networks, Article available at `http://psrg.lcs.mit.edu/6892/handouts/tutbayesheckerman.pdf`, 1995.

[Heckerman *et al.*94] Heckerman, D., Geiger, D., and Chickering, M., Learning Bayesian networks: The combination of knowledge and statistical data, Technical Report MSR-TR-94-09, Redmond, WA: Microsoft Research, 1994.

[Hedenfalk *et al.*01] Hedenfalk, I., Duggan, D., Chen, Y., Radmacher, M., Bittner, M., Simon, R., Meltzer, P., Gusterson, B., Esteller, M., Kallioniemi, O., Wilfond, B., Borg, A., and Trent, J., Gene-expression profiles in hereditary breast cancer, *The New England Journal of Medicine*, vol. 344, no. 8, pp. 539–548, 2001.

[Hiemstra96] Hiemstra, Y., Applying neural networks and genetic algorithms to tactical asset allocation, *NeuroVest Journal*, May/June, 1996.

[Ho98] Ho, T. K., The random subspace method for constructing decision forests, *IEEE Transactions on Pattern Analysis and Machine Intelligence*, vol. 20, no. 8, pp. 832–844, 1998.

[Holland75] Holland. J., *Adaptation in Natural and Artificial Systems*, The University of Michigan Press, 1975.

[Hong and Cho04] Hong, J.-H. and Cho, S. B., Lymphoma cancer classification using genetic programming with SNR features, in *Proc. of Genetic Programming*

7th European Conference, EuroGP 2004, vol. 3003 of *LNCS*, pp. 78–88, Coimbra, Portugal, Springer-Verlag, 2004.

[Hong and Cho06] Hong, J.-H. and Cho, S.-B., The classification of cancer based on DNA microarray data that uses diverse ensemble genetic programming, *Artificial Intelligence in Medicine*, vol. 36, no. 1, pp. 43–58, 2006.

[Hübner *et al.*94] Hübner, U., Weiss, C.-O., Abraham, N. B., and Tang, D., Lorenz-like chaos in NH3-FIR lasers, *Time Series Prediction: Forecasting the Future and Understanding the Past*, Weigend, A. S. and Gershenfeld, N. A. (eds.), pp. 73-104, Addison-Wesley, 1994.

[Hulse *et al.*07] Hulse, J. V., Khoshgoftaar, T. M., and Napolitano, A., Experimental perspectives on learning from imbalanced data, in *Proc. of the 24th International Conference on Machine Learning (ICML07)*, pp. 935–942, New York, NY, ACM, 2007.

[Iba99] Iba, H., Bagging, boosting, and bloating in genetic programming, in *Proc. of the Genetic and Evolutionary Computation Conference (GECCO1999)*, Banzhaf, W., Daida, J., Eiben, A. E., Garzon, M. H., Honavar, V., Jakiela, M., and Smith, R. E. (eds.), pp. 1053–1060, Orlando, Florida, Morgan Kaufmann, 1999.

[Iba02] Iba, H. and Mimura, A., Inference of a gene regulatory network by means of interactive evolutionary computing, *Information Sciences*, vol. 145, no. 3-4, pp. 225–236, 2002.

[Iba *et al.*93] Iba, H., Kurita, T., deGaris, H., and Sato, T., System identification using structured genetic algorithms, in *Proc. of the 5th International Joint Conference on Genetic Algorithms (ICGA93)*, Forrest, S. (ed.), University of Illinois at Urbana-Champaign, pp. 279–286, Morgan Kaufmann, 1993.

[Iba *et al.*94a] Iba, H., deGaris, H. and Sato, T., Genetic programming using a minimum description length principle, *Advances in Genetic Programming*, Kinnear, K. (ed.), pp. 265–284, MIT Press, 1994.

[Iba and Sato94b] Iba, H. and Sato, T., Genetic programming with local hill-climbing, *Parallel Problem Solving from Nature III (PPSN III)*, Davidor, Y., Schwefel, H.-P., and Männer, R. (eds.), vol. 866 of *LNCS*, Jerusalem, 9-14 October, pp. 302–411, Springer-Verlag, 1994.

[Iba *et al.*94c] Iba, H., deGaris, H., and Sato, T., System identification approach to genetic programming, in *Proc. of IEEE World Congress on Computational Intelligence (WCCI1994)*, pp. 401–406, IEEE Press, 1994.

[Iba *et al.*95] Iba, H., deGaris, H., and Sato, T., Temporal data processing using genetic programming, in *Proc. of the 6th International Conference on Genetic Algorithms (ICGA95)*, Eshelman, L. J. (ed.), Pittsburgh, PA, 15-19 July, pp. 279–286, Morgan Kaufmann, 1995.

[Iba *et al.*96a] Iba, H., deGaris, H., and Sato, T., Numerical Approach to genetic programming for system identification, *Evolutionary Computation*, vol. 3, no. 4, pp. 417–452, 1996.

[Iba *et al.*96b] Iba, H., deGaris, H., and Sato, T., Extending genetic programming with recombinative guidance, *Advances in Genetic Programming 2*, Angeline, P. and Kinnear, K. (eds.), MIT Press, 1996.

[Iba96c] Iba, H., Random tree generation for genetic programming, *Parallel Problem Solving from Nature IV (PPSN IV)*, Voigt, H.-M., Ebeling, W., Rechenberg, I., and Schwefel, H.-P. (eds.), vol. 1411 of *LNCS*, pp. 144–153, Springer-Verlag, 1996.

[Iba *et al.*04] Iba, H., Tohge, T., and Inoue, Y., Cooperative transportation by humanoid robots: Solving piano movers' problem, *International Journal of Hybrid Intelligent Systems*, vol. 1, no. 4, pp. 189–201, 2004.

[Ikeda79] Ikeda, K., Multiple-valued stationary state and its instability of the transmitted light by a ring cavity system, *Optics Communications* vol. 30, pp. 257–261, 1979.

[Imamura *et al.*03] Imamura, K., Soule, T., Heckendorn, R. B., and Foster, J. A., Behavioral diversity and a probabilistically optimal GP ensemble, *Genetic Programming* *and Evolvable Machines*, vol. 4, no. 3, pp. 235–253, 2003.

[Inoue *et al.*07] Inoue, Y., Tohge, T., and Iba, H., Cooperative transportation system for humanoid robots using simulation-based learning, *Applied Soft Computing*, vol. 7, no. 1, pp. 115–125, 2007.

[Ivakhnenko71] Ivakhnenko, A. G., Polynomial theory of complex systems, *IEEE Transactions Systems, Man and Cybernetics*, vol. SMC-1, no. 4, 1971.

[Jo and Japkowicz04] Jo, T. and Japkowicz, N., Class imbalances versus small disjuncts, *ACM SIGKDD Explorations Newsletter*, vol. 6, no. 1, pp. 40–49, 2004.

[Keith and Martin94] Keith, M. J. and Martin, M. C., Genetic programming in C++: Implementation issues, *Advances in Genetic Programming*, Kinnear, K. (ed.), pp. 285–310, MIT Press, 1994.

[Keller *et al.*00] Keller, A., Schummer, M., Hood, L., and Ruzzo, W. L., Bayesian classification of DNA array expression data, Technical Report UW-CSE-2000-08-01, Department of Computer Science and Engineering, University of Washington, Seattle, 2000.

[Kim *et al.*04] Kim, Y.-H., Lee, S.-Y., and Moon, B.-R., A genetic approach for gene selection on microarray expression data, in *Proc. of the 6th Genetic and Evolutionary Computation Conference (GECCO2004)*, vol. 3103 of *LNCS*, pp. 346–355, Springer-Verlag, 2004.

[Kirchhofer *et al.*05] Kirchhofer, D., Peek, M., Lipari, M., Billeci, K., Fan, B., and Moran, P., Hepsin activates prohepatocyte growth factor and is inhibited by hepatocyte growth factor activator inhibitor-1B (HAI-1B) and HAI-2, *FEBS Letters*, vol. 579, no. 9, pp. 1945–1950, 2005.

[Kirkpatrick *et al.*83] Kirkpatrick, S., Gelatt, C. D. and Vecchi, M. P., Optimization by simulated annealing, *Science*, vol. 220, pp. 671–680, 1983.

[Kitano90] Kitano, H., Designing neural networks using genetic algorithms with graph generation system, *Complex Systems*, vol. 4, pp. 461-476, 1990.

[Kohavi95] Kohavi, R., A study of cross-validation and boot-strap for accuracy estimation and model selection, in *Proc. of the International Joint Conference on Artificial Intelligence (IJCAI)*, Taylor & Francis, 1995.

[Kohavi and John97] Kohavi, R. and John, G. H., Wrappers for feature subset selection, *Artificial Intelligence*, vol. 97, no. 1–2, pp. 273–324, 1997.

[Koza90] Koza, J., Genetic programming: A paradigm for genetically breeding populations of computer programs to solve problems, Report No. STAN-CS-90-1314, Dept. of Computer Science, Stanford Univ., 1990.

[Koza91] Koza, J., Evolution of subsumption using genetic programming, in *Towards a Practice of Autonomous Systems: Proc. of the 1st European Conference on Artificial Life (ECAl91)*, pp. 110–119, MIT Press, 1992.

[Koza92] Koza, J., *Genetic Programming, On the Programming of Computers by means of Natural Selection*, MIT Press, 1992.

[Koza94] Koza, J., *Genetic Programming II: Automatic Discovery of Reusable Programs*, MIT Press, 1994.

[Koza *et al.*99] Koza, J., Bennett, F. H., Andre, D., and Keane, M. A., *Genetic Programming III: Darwinian Invention and Problem Solving*, Morgan Kaufmann, 1999.

[Koza *et al.*03] Koza, J., Keane, M. A., Streeter, M. J., Mydlowec, W., Yu, J., and Lanza, G., *Genetic Programming IV: Routine Human-Competitive Machine Intelligence*, Kluwer Academic Publishers, 2003.

[Kubat and Matwin97] Kubat, M. and Matwin, S., Addressing the curse of imbalanced data sets: One-sided selection, in *Proc. of the 14th International Conference on Machine Learning (ICML97)*, pp. 179–186, Morgan Kaufmann, 1997.

[Kuncheva and Whitaker01] Kuncheva, L. I. and Whitaker, C. J., Ten measures of diversity in classifier ensembles: Limits for two classifiers, in *Proc. of IEE Workshop on Intelligent Sensor Processing*, pp. 1–10, 2001.

[Kutza96] Kutza, K., Neural networks at your fingertips, available at `http://www.geocities.com/CapeCanaveral/1624/`, 1996.

[Langdon00] Langdon, W. B., Size fair and homologous tree crossovers for tree genetic programming, *Genetic Programming and Evolvable Machines*, vol. 1, no. 1–2, pp. 95–119, 2000.

[Langdon and Poli97] Langdon, W. B. and Poli, R., An analysis of the MAX problem in genetic programming, in *Genetic Programming 1997: Proceedings of the Second Annual Conference*, Koza, J. R., Deb, D., Dorigo, M., Fogel, D. B., Garzon, M., Iba, H., and Riolo, R. L. (eds.), pp. 222–230, Stanford University, CA, Morgan Kaufmann, 1997.

[Langdon and Poli02] Langdon, W. B. and Poli, R., *Foundations of Genetic Programming*, Springer, 2002.

[Langdon and Buxton04] Langdon, W. B. and Buxton, B. F., Genetic programming for mining DNA chip data from cancer patients, *Genetic Programming and Evolvable Machines*, vol. 5, no. 3, pp. 251–257, 2004.

[Langley and Zytkow89] Langley, P. and Zytkow, J. M., Data-driven approaches to empirical discovery, *Artificial Intelligence*, vol. 40, pp. 283–312, 1989.

[Larrañaga and Lozano01] Larrañaga, P. and Lozano, J., *Estimation of Distribution Algorithms: A New Tool for Evolutionary Optimization*, Kluwer Academic Publishers, Boston, MA, 2001.

[Li et al.06a] Li, B.-Y., Peng, J., Chen, Y.-Q., and Jin, Y.-Q., Classifying unbalanced pattern groups by training neural network, in *Proc. of Advances in Neural Networks (ISNN2006)*, vol. 3972 of *LNCS*, Springer-Verlag, 2006.

[Li et al.06b] Li, G.-Z., Yang, J., Ye, C.-Z., and Geng, D.-Y., Degree prediction of malignancy in brain glioma using support vector machines, *Computers in Biology and Medicine*, vol. 36, pp. 313–325, 2006.

[Li et al.04] Li, L., Umbach, D. M., Terry, P., and Taylor, J. A., Application of the GA/KNN method to SELDI proteomics data, *Bioinformatics*, vol. 20, no. 10, pp. 1638–1640, 2004.

[Lim09] Lim, M.-H. (ed.), Memetic Computing, http://www.springer.com/engineering/journal/12293, Springer, 2009.

[Ling and Li98] Ling, C. and Li, C., Data mining for direct marketing: Problems and solutions, in *Proc. of the 4th ACM SIGKDD International Conference on Knowledge Discovery and Data Mining*, pp. 73–79, New York, NY, AAAI Press, 1998.

[Liu and Iba01] Liu, J. and Iba, H., Selecting informative genes with parallel genetic algorithms in tissue classification, *Genome Informatics*, pp. 14–23, 2001.

[Liu and Iba02] Liu, J. and Iba, H., Selecting informative genes using a multiobjective evolutionary algorithm, in *Proc. of the 2002 IEEE World Congress on Computational Intelligence (WCCI2002)*, pp. 297–302, IEEE Press, 2002.

[Liu *et al.*05] Liu, J. J., Cutler, G., Li, W., Pan, Z., Peng, S., Hoey, T., Chen, L., and Ling, X. B., Multiclass cancer classification and biomarker discovery using GA-based algorithms, *Bioinformatics*, vol. 21, no. 11, pp. 2691–2697, 2005.

[Lorenz63] Lorenz, E. N., Deterministic non-periodic flow, *Journal of the Atmospheric Sciences*, vol. 20, p. 130, 1963.

[MacKay95] MacKay, D. J. C., Probable networks and plausible predictions: A review of practical Bayesian methods for supervised neural networks, *Network: Computation in Neural Systems*, vol. 6, no. 3, pp. 469–505, 1995.

[Mackey and Glass77] Mackey, M. C. and Glass, L., Oscillation and chaos in physiological control systems, *Science*, vol. 197, pp. 287–107, 1977.

[Matsuzaki *et al.*05] Matsuzaki, T., Miyao, Y., and Tsujii, J., Probabilistic CFG with latent annotations, in *Proc. of the 43rd Meeting of the Association for Computational Linguistics (ACL)*, pp. 75–82, Morgan Kaufmann, 2005.

[Matthews75] Matthews, B., Comparison of the predicted and observed secondary structure of T4 phage lysozyme, *Biochemica et Biophysica Acta.*, vol. 405, pp. 442–451, 1975.

[McDonnell and Waagen94] MacDonnell, J. R. and Waagen, D. Evolving recurrent perceptrons for time-series modeling, *IEEE Transactions on Neural Networks*, vol. 5, no. 1, pp. 24–38, 1994.

[Mitchell97] Mitchell, T., *Machine Learning*, McGraw Hill, 1997.

[Mitchell *et al.*92] Mitchell, M., Forrest, S., and Holland, J. H., The royal road for genetic algorithms: Fitness landscapes and GA performance, in *Towards a Practice of Autonomous Systems: Proc. of the 1st European Conference on Artificial Life (ECAL91)*, pp. 245–254, MIT Press, 1992.

[Mitra *et al.*06] Mitra, A. P., Almal, A. A., George, B., Fry, D. W., Lenehan, P. F., Pagliarulo, V., Cote, R. J., Datar, R. H., and Worzel, W. P., The use of genetic programming in the analysis of quantitative gene expression profiles for identification of nodal status in bladder cancer, *BMC Cancer*, vol. 6, no. 159, 2006.

[Mühlenbein and Schlierkamp95] Mühlenbein, H. and Schlierkamp, D., Predictive models for the breeder genetic algorithm: I, *Evolutionary Computation*, vol. 1, no. 1, pp. 25–49, 1995.

[Mühlenbein and Paaß96] Mühlenbein, H. and Paaß, G., From recombination of genes to the estimation of distribution I. Binary parameters, *Parallel Problem Solving from Nature IV (PPSN IV)*, Voigt, H.-M., Ebeling, W., Rechenberg, I., and Schwefel, H.-P. (eds.), vol. 1411 of *LNCS*, pp. 178–187, Springer-Verlag, 1996.

[Muni *et al.*04] Muni, D. P., Pal, N. R., and Das, J., A novel approach to design classifiers using genetic programming, *IEEE Transaction on Evolutionary Computation*, vol. 8, no. 2, pp. 183–196, 2004.

[Murphey *et al.*04] Murphey, Y. L., Guo, H., and Feldkamp, L. A., Neural learning from unbalanced data, *Applied Intelligence*, vol. 21, no. 2, pp. 117–128, 2004.

[Myers,1990] Myers, R. H., *Classical and Modern Regression with Applications*, Duxbury Press, 1990.

[Newman *et al.*98] Newman, D. J., Hettich, S., Blake, C. L., and Merz, C. J., UCI repository of machine learning databases, `http://archive.ics.uci.edu/ml/`, 1998.

[Nikolaev and Iba06] Nikolaev, N. and Iba, H., *Adaptive Learning of Polynomial Networks Genetic Programming, Back-propagation and Bayesian Methods*, Springer, 2006.

[Nilsson80] Nilsson, N. J., *Principles of Artificial Intelligence*, Tioga Publishing Company, 1980.

[Nilsson98] Nilsson, N. J., *Artificial Intelligence: A New Synthesis*, Morgan Kaufmann, 1998.

[Nordin94] Nordin, P., A compiling genetic programming system that directly manipulates the machine code. *Advances in Genetic Programming*, Kinnear, K. (ed.), pp. 311–331, MIT Press, 1994.

[Nutt *et al.*03] Nutt, C., Mani, D., Betensky, R., Tamayo, P., Cairncross, J., Ladd, C., Pohl, U., Hartmann, C., McLaughlin, M., Batchelor, T. T., Black, P., von Deimling, A., Pomeroy, S., Golub, T., and Louis, D., Gene expression-based classification of malignant gliomas correlates better with survival than histological classification, *Cancer Research*, vol. 63, no. 7, pp. 1602–1607, 2003.

[Oakley94] Oakley, H., Two scientific applications of genetic programming: Stack filters and non-linear equation fitting to chaotic data, *Advances in Genetic Programming*, Kinnear, K. (ed.), MIT Press, 1994.

[Onn *et al.*04] Onn, A., Correa, A. M., Gilcrease, M., Isobe, T., Massarelli, E., Bucana, C. D., O'Reilly, M. S., Hong, W. K., Fidler, I. J., Putnam, J. B., and Herbst, R. S., Synchronous overexpression of epidermal growth factor receptor and HER2-NEU protein is a predictor of poor outcome in patients with stage I non-small cell lung cancer, *Clinical Cancer Research*, vol. 10, pp. 136–143, 2004.

[Ooi and Tan03] Ooi, C. H. and Tan, P., Genetic algorithms applied to multi-class prediction for the analysis of gene expression data, *Bioinformatics*, vol. 19, no. 1, pp. 37–44, 2003.

[O'Reilly *et al.*05] O'Reilly, U.-M., Yu, T., Riolo, R., and Worzel, B. (Eds.), *Genetic Programming Theory and Practice II*, Springer, 2005.

[Pan *et al.*04] Pan, F., Wang, B., Hu, X., and Perrizo, W., Comprehensive vertical sample-based KNN/LSVM classification for gene expression analysis, *Journal of*

Biomedical Informatics, vol. 37, no. 4, pp. 240–248, 2004.

[Park *et al.*01]   Park, P., Pagano, M., and Bonnetti, M., A non-parametric scoring algorithm for identifying informative genes from microarray data, in *Proc. of Pacific Symposium on Bioinformatics (PSB)*, vol. 6, pp. 30–41, 2001.

[Paul and Iba03a]   Paul, T. K. and Iba, H., Linear and combinatorial optimizations by estimation of distribution algorithms, in *Proc. of the 9th MPS Symposium on Evolutionary Computation*, pp. 99–106, The Information Processing Society of Japan (IPSJ), Article available at http://www.iba.t.u-tokyo.ac.jp/english/EDA.htm, 2003.

[Paul and Iba03b]   Paul, T. K. and Iba, H., Optimization in continuous domain by real-coded estimation of distribution algorithm, *Design and application of hybrid intelligent systems*, pp. 262–271, IOS Press, Amsterdam, Netherlands, 2003.

[Paul and Iba04a]   Paul, T. K. and Iba, H., Identification of informative genes for molecular classification using probabilistic model building genetic algorithm, in *Proc. of the 6th Genetic and Evolutionary Computation Conference 2004 (GECCO2004)*, vol. 3103 of *LNCS*, pp. 414–425, Springer-Verlag, 2004.

[Paul and Iba04b]   Paul, T. K. and Iba, H., Selection of the most useful subset of genes for gene expression-based classification, in *Proc. of the 2004 Congress on Evolutionary Computation (CEC2004)*, Portland, Oregon, pp. 2076–2083, IEEE Computational Intelligence Society, IEEE Press, 2004.

[Paul and Iba05]   Paul, T. K. and Iba, H., Gene selection for classification of cancers using probabilistic model building genetic algorithm, *BioSystems*, vol. 82, no. 3, pp. 208–225, 2005.

[Paul and Iba06a]   Paul, T. K. and Iba, H., Classification of scleroderma and normal biopsy data and identification of possible biomarkers of the disease, in *Proc. of the IEEE Symposium on Computational Intelligence in Bioinformatics and Computational Biology 2006 (CIBCB2006)*, pp. 306–311, Toronto, Ontario,

Canada, IEEE Computational Intelligence Society, 2006.

[Paul and Iba07] Paul, T. K. and Iba, H., Genetic programming for classifying cancer data and controlling humanoid robots, *Genetic Programming Theory and Practice IV*, Riolo, R. L., Terence, S., and Worzel, B. (eds.), pp. 41–60, Kluwer Academic Publishers, 2007.

[Paul and Iba08] Paul, T. K. and Iba, H., Prediction of cancer class with majority voting genetic programming classifier using gene expression data, *IEEE/ACM Transactions on Computational Biology and Bioinformatics*, Preprint on IEEE Computer Society Digital Library, 11 June 2008, http://doi.ieeecomputersociety.org/10.1109/TCBB.2007.70245.

[Paul *et al.*06] Paul, T. K., Hasegawa, Y., and Iba, H., Classification of gene expression data by majority voting genetic programming classifier, in *Proc. of the 2006 IEEE Congress on Evolutionary Computation (CEC2006)*, Vancouver, BC, Canada, pp. 2521–2528, IEEE Computational Intelligence Society, IEEE Press, 2006.

[Paul *et al.*08a] Paul, T. K., Ueno, K., Iwata, K., Hayashi, T., and Honda, N., Genetic algorithm based methods for identification of health risk factors aimed at preventing metabolic syndrome, in *Proc. of the 7th International Conference on Simulated Evolution and Learning (SEAL'08)*, vol. 5361 of LNCS, pp. 210–219, Springer-Verlag, 2008.

[Paul *et al.*08b] Paul, T. K., Ueno, K., Iwata, K., Hayashi, T., and Honda, N., Risk prediction and risk factors identification from imbalanced data with RPMBGA+, in *Proc. of the 10th Genetic and Evolutionary Computation Conference (GECCO2008)*, pp. 2193–2198, ACM Press, 2008.

[Pavlidis *et al.*01] Pavlidis, P., Weston, J., Cai, J., and Grundy, W. N., Gene functional classification from heterogeneous data, in *Proc. of the 5th International Conference on Computational Modelcular Biology (RE-COMB)*, pp. 249–255, Assn for Computing Machinery, 2001.

[Pelikan *et al.*99a] Pelikan, M., Goldberg, D., and Lobo, F., A survey of optimizations by building and using probabilistic models, Technical Report, Illigal Report 99018, Illinois Genetic Algorithms Laboratory, University of Illinois at Urbana-Champaign, 1999.

[Pelikan *et al.*99b] Pelikan, M., Goldberg, D. E., and Cantú-Paz, E., BOA: The Bayesian optimization algorithm, in *Proc. of the Genetic and Evolutionary Computation Conference (GECCO1999)*, Banzhaf, W., Daida, J., Eiben, A. E., Garzon, M. H., Honavar, V., Jakiela, M., and Smith, R. E. (eds.), pp. 525–532, Orlando, Florida, Morgan Kaufmann, 1999.

[Pena *et al.*05] Peña, J. M., Lozano, J. A., and Larrañaga,P., Globally multimodal problem optimization via an estimation of distribution algorithm based on unsupervised learning of Bayesian networks, *Evolutionary Computation*, vol. 13, no. 1, pp. 43–66, 2005.

[Poggio and Girosi90] Poggio, T. and Girosi, F., Networks for Approximation and Learning, in *Proc. of the IEEE*, vol. 78, no. 9, pp. 1481–1497, 1990.

[Poli and McPhee08] Poli, R. and McPhee, N. F., A linear estimation-of-distribution GP system, in *Proc. of Genetic Programming 11th European Conference, EuroGP 2008*, vol. 4971 of *LNCS*, pp. 206–217, Naples, Italy, Springer-Verlag, 2008.

[Polikar06] Polikar, R., Ensemble based systems in decision making, *IEEE Circuits and Systems Magazine*, vol. 6, no. 3, pp. 21–45, 2006.

[Potvin *et al.*04] Potvin, J.-Y., Soriano, P., and Vallee, M., Generating trading rules on the stock markets, *Computer & Operations Research*, vol. 31, pp. 1033–1047, 2004.

[Press *et al.*88] Press, W. H., Flannery, B. P., Teukolsky, S. A., and Vetterling, W. T., *Numerical Recipes in C*, Cambridge University Press, 1988.

[Punch98] Punch, W. F., How effective are multiple populations in genetic programming, in *Genetic Programming 1998: Proc. of the 3rd Annual Conference*, University of Wisconsin, Madison, Wisconsin, 22-25 July, pp. 308–313, Morgan Kaufmann, 1998.

[Quinlan93] Quinlan, J., *C4.5:Programs for Machine Learning*, Morgan Kaufmann, 1993.

[Ramaswamy *et al.*01] Ramaswamy, S., Tamayo, P., Rifkin, R., Mukherjee, S., Yeang, C.-H., Angelo, M., Ladd, C., Reich, M., Latulippe, E., Mesirov, J., Poggio, T., Gerald, W., Loda, M., Lander, E., and Golub, T., Multiclass cancer diagnosis using tumor gene expression signatures, *Proceedings of National Academy of Science*, vol. 98, no. 26, pp. 15149–15154, 2001.

[Ratle and Sebag01] Ratle, A. and Sebag, M., Avoiding the bloat with probabilistic grammar-guided genetic programming, in *Proc. of the Artificial Evolution 5th International Conference, Evolution Artificielle, EA 2001*, vol. 2310 of *LNCS*, Creusot, France, 29-31 October, pp. 255–266, Springer-Verlag, 2001.

[Riolo and Worzel03] Riolo, R. L. and Worzel, B. (eds.), *Genetic Programming Theory and Practice*, Kluwer Academic Publishers, 2003.

[Rissanen78] Rissanen, J., Modeling by shortest data description, *Automatica*, pp. 465–471, 1978.

[Rowland03] Rowland, J. J., Interpreting analytical spectra with evolutionary computation, *Evolutionary Computation in Bioinformatics*, Fogel, G. B. and Corne, D. W. (eds.), pp. 341–365, Morgan Kaufmann, 2003.

[Rumelhart *et al.*86] Rumelhart, D. E., Hinton, G. E., and Williams, R. J., Learning internal representations by error propagation, *Parallel Distributed Processing: Explorations in the Microstructure of Cognition*, vol. 1, Rumelhart, D. E. et al. (eds.), pp. 318-362, MIT Press, 1986.

[Sałustowicz and Schmidhuber97a] Sałustowicz, R. P. and Schmidhuber, J., Probabilistic incremental program evolution, *Evolutionary Computation*, vol. 5, no. 2, pp. 123–141, 1997.

[Sałustowicz and Schmidhuber97b] Sałustowicz, R. P. and Schmidhuber, J., Probabilistic incremental program evolution: Stochastic search through program space, in *Proc. of the 9th European Conference on Machine Learning (ECML97)*, vol. 1224 of *LNAI*, pp. 213–220, Springer-Verlag, 1997.

[Santana05] Santana, R., Estimation of distribution algorithms with Kikuchi approximations, *Evolutionary Computation*, vol. 13, no. 1, pp. 67–97, 2005.

[Sastry and Goldberg03] Sastry, K. and Goldberg, D. E., Probabilistic model building and competent genetic programming, *Genetic Programming Theory and Practice*, Riolo, R. L. and Worzel, B. (eds.), pp. 205–220, Kluwer Academic Publishers, 2003.

[Schaffer and Morishima87] Schaffer, J. D. and Morishima, A., An adaptive crossover distribution mechanism for genetic algorithms, in *Proc. of the 2nd International Joint Conference on Genetic Algorithms (ICGA87)*, Grefenstette, J. J. (ed.), pp. 36–40, Lawrence Erlbaum Associates, 1987.

[Schapire99] Schapire, R., A brief introduction to boosting, in *Proc. of the 16th International Joint Conference on Artificial Intelligence*, pp. 1401–1406, Morgan Kaufmann, 1999.

[Schwarz78] Schwarz, G., Estimating the dimension of a model, *Annals of Statistics*, vol. 6, no. 2, pp. 461–464, 1978.

[Schwefel95] Schwefel, H. P., *Evolution and Optimum Seeking*, John Wiley & Sons, 1995.

[Shan et al.03] Shan, Y., McKay, R. I., Abbass, H. A., and Essam, D., Program evolution with explicit learning: A new framework for program automatic synthesis, in *Proc. of the 2003 Congress on Evolutionary Computation (CEC2003)*, Canberra, Australia, 8-12 December, pp. 1639–1646, IEEE Computational Intelligence Society, IEEE Press, 2003.

[Shan et al.04] Shan, Y., McKay, R. I., Baxter, R., Abbass, H., Essam, D., and Hoai, N. X., Grammar model-based program evolution, in *Proc. of the 2004 IEEE Congress on Evolutionary Computation (CEC2004)*, Portland, Oregon, 20-23 June, pp. 478–485, IEEE Computational Intelligence Society, 2004.

[Shen and Tan06] Shen, L. and Tan, E. C., A generalized output-coding scheme with SVM for multiclass microarray classification, in *Proc. of the 4th Asia-Pacific Bioinformatics Conference*, pp. 179–186, 2006.

[Shin et al.07] Shin, J., Kang, M., McKay, R. I., Nguyen, X., Hoang, T.-H., Mori, N., and Essam, D., Analysing the regularity of genomes using compression and expression simplification, in *Proc. of Genetic*

Programming 10th European Conference, EuroGP 2007, vol. 4445 of *LNCS*, pp. 251–260, Valencia, Spain, Springer-Verlag, 2007.

[Singh *et al.*02]   Singh, D., Febbo, P., Ross, K., Jackson, D., Manola, J., Ladd, C., Tamayo, P., Renshaw, A., D'Amico, A., Richie, J., Lander, E., Loda, M., Kantoff, P., Golub, T., and Sellers, W., Gene expression correlates of clinical prostate cancer behavior, *Cancer Cell*, vol. 1, no. 2, pp. 203–209, 2002.

[Slonim *et al.*00]   Slonim, D., Tamayo, P., Mesirov, J., Golub, T., and Lander, E., Class prediction and discovery using gene expression data, in *Proc. of the 4th Annual International Conference on Computational Molecular Biology*, pp. 263–272, Tokyo, Japan, 2000.

[Smirnov01]   Smirnov, E. N., *Conjunctive and Disjunctive Version Spaces with Instance-based Boundary Sets*, Shaker Publishing, 2001.

[Spiegel75]   Spiegel, M. R., *Theory and Problems of Statistics*, McGraw-Hill, 1975.

[Stolcke94]   Stolcke, A., Bayesian learning of probabilistic language models, PhD thesis, University of California, Berkeley, CA, 1994.

[Sun90]   Sun, G. Z., Chen, H. H., Giles, C. L., Lee, Y. C., and Chen, D., Connectionist pushdown automata that learn context-free grammars, in *Proc. of the International Joint Conference on Neural Networks*, pp. 577–580, Lawrence Erlbaum Associates, 1990.

[Sun *et al.*06]   Sun, Y., Kamel, M. S., and Wang, Y., Boosting for learning multiple classes with imbalanced class distribution, in *Proc. of the 6th International Conference on Data Mining (ICDM06)*, pp. 592–602, Washington, DC, IEEE Computer Society, 2006.

[Tackett93]   Tackett, W. A., Genetic programming for feature discovery and image discrimination, in *Proc. of the 5th International Conference on Genetic Algorithms (ICGA93)*, Forrest, S. (ed.), University of Illinois at Urbana-Champaign, pp. 303–309, Morgan Kaufmann, 1993.

[Takagi01]   Takagi, H., Interactive evolutionary computation: Fusion of the capacities of EC optimization and human evaluation, in *Proc. of the IEEE*, vol. 89, no. 9, pp. 1275–1296, 2001.

[Tan *et al.*02] Tan, K. C., Tay, A., Lee, T. H., and Heng, C. M., Mining multiple comprehensible classification rules using genetic programming, in *Proc. of the 2002 IEEE World Congress on Computational Intelligence (WCCI2002)*, pp. 1302–1307, IEEE Press, 2002.

[Tanji and Iba09] Tanji, M. and Iba, H., Program Optimization by Random Tree Sampling, in *Proc. of the 11th Genetic and Evolutionary Computation Conference (GECCO2009)*, ACM Press, 2008.

[Tanev05] Tanev, I., Incorporating learning probabilistic context-sensitive grammar in genetic programming for efficient evolution and adaptation of Snakebot, in *Proc. of Genetic Programming 8th European Conference, EuroGP 2005*, vol. 3447 of *LNCS*, pp. 155–166, Lausanne, Switzerland, Springer-Verlag, 2005.

[Tanev04] Tanev, I., Implications of incorporating learning probabilistic context-sensitive grammar in genetic programming on evolvability of adaptive locomotion gaits of Snakebot,in *GECCO 2004 Workshop Proceedings*, Poli, R. et. al (eds.), Seattle, Washington, 26-30 June, 2004.

[Teller and Veloso1996] Teller, A. and Veloso, M., PADO: A new learning architecture for object recognition, *Symbolic Visual Learning*, Ikeuchi, K. and Veloso, M. (eds.), pp. 81–116, Oxford University Press, 1996.

[Tenorio *et al.*90] Tenorio, M. F. and Lee, W., Self-organizing network for optimum supervised learning, *IEEE Transactions on Neural Networks*, vol. 1, no. 1, pp. 100–109, 1990.

[Tokui and Iba99] Tokui, N. and Iba, H., Empirical and statistical analysis of genetic programming with linear genome, in *Proc. of IEEE International Conference on Systems, Man and Cybernetics (SMC99)*, vol. III, pp. 610–615, IEEE Press, 1999.

[Tomita82] Tomata, M., Dynamic construction of finite automata from examples using hill-climbing, in *Proc.*

of the 4th International Cognitive Science Conference, 1982.

[Unemi99] Unemi, T., SBART2.4: Breeding 2D CG images and movies, and creating a type of collage, In *Proc. of the 3rd International Conference on Knowledge-based Intelligent Information Engineering Systems*, pp. 288–291, Adelaide, Australia, 1999.

[Vapnik95] Vapnik, V., *The Nature of Statistical Learning Theory*, Springer-Verlag, New York, 1995.

[Wang *et al.*03] Wang, H., Wang, H., Shen, W., Huang, H., Hu, L., Ramdas, L., Zhou, Y., Liao, W., Fuller, G., and Zhang, W., Insulin-like growth factor binding protein 2 enhances glioblastoma invasion by activating invasion-enhancing genes, *Cancer Research*, vol. 63, no. 15, pp. 4315–4321, 2003.

[Watrous and Kuhn92] Watrous, R. L. and Kuhn, G. M., Induction of finite-state languages using second-order recurrent networks, *Neural Computation*, vol. 4, 1992.

[Whigham95] Whigham, P. A., Grammatically-based genetic programming, in *Proc. of the Workshop on Genetic Programming : From Theory to Real-World Applications*, Rosca, J. P. (ed.), pp. 44–41, Tahoe City, California, 1995.

[Whigham96] Whigham, P. A., Search bias, language bias, and genetic programming, in *Genetic Programming 1996: Proceedings of the First Annual Conference*, Koza, J., Goldberg, D. E., Fogel, D. B., and Riolo, R. L. (eds.), Stanford University, CA, 28–31 July, pp. 230–237, MIT Press, 1996.

[Williams and Zipser89] Williams, R. J. and Zipser, D., Experimental analysis of the real-time recurrent learning algorithm, *Connection Science*, vol. 1, no. 1, 1989.

[Wilson87] Wilson, S. W., Classifier systems and the animat problem, *Machine Learning*, vol. 2, no. 3, 1987.

[Wineberg and Oppacher94] Wineberg, M. and Oppacher, F., A representation scheme to perform program induction in a canonical genetic algorithm, *Parallel Problem Solving from Nature III (PPSN III)*, Davidor, Y., Schwefel, H.-P., and Männer, R. (eds.), vol. 866 of *LNCS*, Jerusalem, 9-14 October, pp. 292–301, Springer-Verlag, 1994.

318   *References*

[Witten and Frank05]   Witten, I. H. and Frank, E., *Data Mining: Practical machine learning tools and techniques*, , 2nd edition, Morgan Kaufmann, 2005.

[Wray and Green94]   Wray, J. and Green, G. G. R., Calculation of the Volterra kernels of non-linear dynamic systems using an artificial neural networks, *Biological Cybernetics*, vol. 71, no. 3, pp. 187–195, 1994.

[Yanai and Iba03]   Yanai, K. and Iba, H., Estimation of distribution programming based on Bayesian network, in *Proc. of the 2003 Congress on Evolutionary Computation (CEC2003)*, Canberra, Australia, 8-12 December, pp. 1618–1625, IEEE Computational Intelligence Society, IEEE Press, 2003.

[Yanai and Iba04]   Yanai, K. and Iba, H., Program evolution by integrating EDP and GP, in *Proc. of the 6th Genetic and Evolutionary Computation Conference 2004 (GECCO2004)*, vol. 3103 of *LNCS*, pp. 774–785, Springer-Verlag, 2004.

[Yanai and Iba05]   Yanai, K. and Iba, H., Probabilistic distribution models for EDA-based GP, in *Proc. of the 7th Genetic and Evolutionary Computation Conference 2005 (GECCO2005)*, pp. 1775–1776, ACM Press, 2005.

[Yanase and Iba09]   Yanase, T. and Iba, H., Binary Encoding for Prototype Tree of Probabilistic Model Building GP, in *Proc. of the 11th Genetic and Evolutionary Computation Conference (GECCO2009)*, ACM Press, 2008.

[Yu *et al.*06a]   Yu, T., Riolo, R., and Worzel, B. (eds.), *Genetic Programming Theory and Practice III*, Springer-Verlag, 2006.

[Yu *et al.*06b]   Yu, T., Wilkinson, D., and Castellini, A., Applying genetic programming to reservoir history matching problem, *Genetic Programming Theory and Practice IV*, Riolo, R. L., Terence, S., and Worzel, B. (eds.), pp. 187–201, Kluwer Academic Publishers, 2007.

[Yule1900]   Yule, G. U., On the association of attributes in statistics, *Philosophical Transactions of the Royal Society of London*, vol. 194, pp. 257–319, 1900.

[Zhang and Mühlenbein93]   Zhang, B. T. and Mühlenbein, H., Genetic programming of minimal neural networks using Occam's razor, in *Proc. of the 5th International Conference on Genetic Algorithms (ICGA93)*, Forrest, S. (ed.), University of Illinois at Urbana-Champaign, pp. 342–349, Morgan Kaufmann, 1993.

[Zhang and Mühlenbein95]   Zhang, B.-T. and Mühlenbein, H., Balancing accuracy and parsimony in genetic programming, *Evolutionary Computation*, vol. 3, no. 1, pp. 17–38, 1995.

[Zhang *et al.*04]   Zhang, J., Bloedorn, E., Rosen, L., and Venese, D., Learning rules from highly unbalanced data sets, in *Proc. of the 4th IEEE International Conference on Data Mining (ICDM04)*, pp. 571- 574, 2004.

[Zhang and Smart06]   Zhang, M. and Smart, W., Using Gaussian distribution to construct fitness functions in genetic programming for multiclass object classification, *Pattern Recognition Letters*, vol. 27, no. 11, pp. 1266–1274, 2006.

[Zhang and Wong08]   Zhang, M. and Wong, P., Genetic programming for medical classification: A program simplification approach, *Genetic Programming and Evolvable Machines*, vol. 9, no. 3, pp. 229–255, 2008.

[Zhang and Bhattacharyya04]   Zhang, Y. and Bhattacharyya, S., Genetic programming in classifying large-scale data: An ensemble method, *Information Sciences*, vol. 163, pp. 85–101, 2004.

Index

k-NN, 126, 129, 170
t-statistics, 132

A-fitness, 143
accuracy-based fitness function, *see* A-fitness
activation polynomial, 109
AdaBoost, 152
ADF, *see* automatic defining function
AI, *see* Artificial Intelligence
ALN, 117
American type option, 99
annotation, 235
ant colony optimization, 231
ANT Simulator, 26
area under ROC curve, *see* AUC
artificial ant, 26
artificial intelligence, 7
artificial intelligence, 1, 5
artificial life, 26
at-the-money, 98
AUC, 144
AUC balanced, *see* AUCB
AUCB, 145, 170
augmented GP, 116
automatic defining function, 45
Ayame, 71

B-fitness, 145
back-propagation, 76, 109
backward probability, 236
bagging, 31, 150
Bayesian automatic programming, 230
Bayesian Dirichlet, 211
Bayesian information criteria, 211
Bayesian network, 208, 213, 221, 224
Bayesian optimization algorithm, 225

benchmark test, 195
biased fitness, *see* B-fitness
bilinear function, 80
binary classification, 168, 172
binary tree, 108, 117
biomarkers, 162
Black–Scholes formula, 99
bloat, 25
Boolean concept formation, 117
Boolean function, 80
boosting, 31, 152
bootstrap aggregating, *see* bagging
box moving task, 33
BP, *see* back-propagation
brain cancer
    data, 160
    genes, 163
breast cancer
    data, 161
    genes, 163
building block, 45, 58, 118, 200, 220

C-fitness, 143
C4.5, 170
call option, 97
CFG, *see* context-free grammar
chaotic time series, 66
CHC selection, 136
child, 8
chromosome, 1, 199
classification, 35, 37, 44, 70
classification statistics, 122, 135, 142
classifier system, 7
classifiers, 126
coding length, 55
complete form, 44, 48, 50
computational cost, 50, 54, 96, 110, 114, 119

321